Metamaterial Inspired Electromagnetic Applications

Balamati Choudhury
Editor

Metamaterial Inspired Electromagnetic Applications

Role of Intelligent Systems

CSIR-NAL

Springer

Editor
Balamati Choudhury
Centre for Electromagnetics
CSIR-National Aerospace Laboratories
Bangalore, Karnataka
India

ISBN 978-981-10-3835-8 ISBN 978-981-10-3836-5 (eBook)
DOI 10.1007/978-981-10-3836-5

Library of Congress Control Number: 2017933062

Printed on acid-free paper

This Springer imprint is published by Springer Nature
The registered company is Springer Nature Singapore Pte Ltd.
The registered company address is: 152 Beach Road, #21-01/04 Gateway East, Singapore 189721, Singapore

It takes a minute to find a special person, an hour to appreciate them, and a day to respect them, but it takes an entire lifetime to forget them. This book is dedicated to the entirely unforgettable Dr. R M Jha!

Preface

The metamaterial science and technology are emerging as a "next higher level" performance contributor, specifically in the electromagnetic community. After experimental verification of negative refractive index metamaterial in 2001, the research community is focusing on application side and an exponential growth has been observed in various metamaterial electromagnetic applications such as antennas, microwave devices, high NA lenses, and radar absorbers. Electromagnetic (EM) designs including metamaterial being complex and nonlinear in nature optimization play an important role. In this book, we focus on role of soft computing technique-based EM computational engines in design and optimization of a wide range of electromagnetic applications. Apart from the theoretical background of metamaterials and soft computing techniques, this contributed volume includes novel electromagnetic applications such as tensor analysis for invisibility cloaking, metamaterial structures for cloaking applications, broadband radar absorbers and antennas.

Chapter 1 gives an insight to background theory of metamaterial and soft computing techniques. The fundamental principle of metamaterial and their properties have been discussed. Toward optimization of these structures, various soft computing algorithms such as *genetic algorithm* (GA), *particle swarm optimization* (PSO), and *bacterial foraging optimization* (BFO) have been described. Various types of metamaterial unit cells and their design and optimization procedures using soft computing techniques have been detailed along with the development of soft computing-based electromagnetic computational engine.

Planar inverted F antennas (PIFA) have wide range of applications in wireless industry. These antennas are also used as wireless sensors in aircraft fuel tanks to check the fuel availability status. Chapter 2 explores the advancements in design optimization of PIFA systems. The literature reveals that the antenna size can significantly be reduced by introducing metamaterial structures. Hence, design optimization of a metamaterial-loaded compact planar inverted F antenna has been carried out and reported systematically. The optimized antenna shows a return loss of −51.57 dB, whereas the conventional PIFA is of −12.893 dB only.

Chapter 3 provides an insight into the scalars, vectors, and tensors in electromagnetics and the significance of their analysis in the aerospace domain. The fundamental mathematical concepts of scalars, vectors, and tensors are being explained, and a thorough investigation on the relevance of these parameters in electromagnetics, cloaking, and various other aerospace applications has been conducted. This chapter connects to the detail design of metamaterial cloaking and optimization of conformal unit cells in the subsequent chapters.

In recent years, the radar cross section (RCS) reduction characteristics of cloaking structures have been widely investigated as it found extensive applications in stealth platform. Chapter 4 deals with the design of ideal circular, cylindrical, and spherical cloaking structures in accordance with transformation optics theory. The simulations are performed by using finite element method (FEM)-based COMSOL Multiphysics software. The performance of the designed ideal spherical cloak is analyzed by comparing the RCS of PEC sphere without cloaking shells and with multilayer cloaking shells. It has been observed that multilayer spherical cloak shows reduced RCS in C band with respect to the PEC sphere. Further, the analysis of particle ray trajectories has been carried out by performing geometric ray tracing.

Conformal metamaterial multilayer analysis for design and simulation of invisibility cloak is a challenging task because of the curvature effects of the platform. Chapter 5 gives the analysis of material properties of various conformal metamaterial unit cell structures patterned on the cloaking layers. Design and simulation of planar and nonplanar metamaterial unit cells have been carried out and reported systematically. The extracted permittivity and permeability characteristics of the conformal metamaterial structures are compared with corresponding planar one. Further analysis of effect of radius of curvature of the platform has been studied and explained briefly.

The advances in artificial engineered materials such as metamaterials have created a wide interest in military aviation scientific community. The researches in modern military aviation are endowed with techniques based on low observable platforms. Design and development of metamaterial-based radar absorbers are one of the cutting edge technologies in this domain. Design and development of two metamaterial-based broadband microwave absorbers have been carried out and described in Chap. 6. Further, the proposed structures have been optimized for absorption in the entire X band, using soft computing-based computational engine. The proposed pentagon-shaped metamaterial structure and novel metamaterial structures have been fabricated, and backscattered signal power from the test samples is measured and compared with backscattered signal power from PEC of same footprint. The measurement result ensures the broadband absorption capability of proposed structures in the X band.

Bangalore, India Balamati Choudhury

Acknowledgements

At the outset, I wish to thank Mr. Jatinder J Jadhav, Director, and Dr. Jatinder Singh, Cluster Chairman, Systems Engineering Cluster of CSIR-National Aerospace Laboratories, Bangalore for sustained support and official permission to write this contributed volume.

I would also like to acknowledge valuable suggestions from Dr. R.U. Nair, Head, Centre for Electromagnetics, CSIR-National Aerospace Laboratories and his invaluable support during the course of writing this book.

Beyond the technical aspects, correction of grammatical error is also a very pertinent area. I would like to extend my thanks to Ms Anusha Eldo for her help in going through the entire book for syntax error editing.

It is our pleasure to acknowledge all the authors of this contributed volume who have completed their research work as project scientists at CSIR-National Aerospace Laboratories. Indeed their work carried out at Centre for Electromagnetics, CSIR-National Aerospace Laboratories has been adapted here as book chapters of this book. We thank them for transferring the necessary copyrights for facilitating the production and circulation of the book in hand.

The publisher's effort plays a very important role in writing a book on scheduled time. Swati Meherishi, Executive Editor at Springer has always been very responsive in this regard and I would like to thank her for all her inputs.

Needless to mention the gratefulness, Balamati owes to her uncle Mr. Bipin Padhy for his blessings and constant encouragement during the course of writing this book.

Balamati Choudhury

Contents

About the Editor

Dr. Balamati Choudhury is a scientist at the Centre for Electromagnetics of the CSIR-National Aerospace Laboratories, Bangalore, India. She received her M. Tech. (ECE) degree from the National Institute of Science and Technology (NIST), India, and Ph.D. (Eng.) degree in microwave engineering from Biju Patnaik University of Technology (BPUT), India, in 2013. From 2006 to 2008, she was a senior lecturer at the Department of Electronics and Communication at the NIST, Orissa, India.

Her research and teaching interests are in the domain of soft computing techniques in electromagnetic design and optimization, computational electromagnetics for aerospace applications, metamaterial design applications, radio frequency (RF), and microwaves. She has contributed to a number of projects, including the development of ray tracing techniques for RF analysis of propagation in an indoor environment, low radar cross section (RCS) design, phased arrays and adaptive arrays, and conformal antennas. She was also the recipient of the CSIR-NAL Young Scientist Award for the year 2013–2014 for her contribution in the area of computational electromagnetics for aerospace applications.

Dr. Balamati has authored or co-authored over 140 scientific research papers and technical reports, five SpringerBriefs, and three book chapters as well as a book entitled: *Soft Computing in Electromagnetics: Methods and Applications*. Dr. Balamati is also an assistant professor of the Academy of Scientific and Innovative Research (AcSIR), New Delhi.

Abbreviations

DNM	Double negative materials
EBG	Electromagnetic band gap
FEM	Finite element methods
FSS	Frequency-selective surface
GSM	Global system for mobile communication
LHM	Left-handed materials
PBG	Photonic band gap
PIFA	Planar inverted F antenna
PSO	Particle swarm optimization
SAR	Specific absorption rate
SRR	Spilt ring resonator
VBA	Visual basic for applications

List of Figures

List of Tables

Chapter 1
Soft Computing for Metamaterial Structures

Balamati Choudhury

1.1 Introduction

Metamaterials are one of the widely used artificially engineered materials in the area of electromagnetic (EM) applications as it exhibits certain properties that are not seen in naturally occurring materials. The most exciting property of metamaterials is the ability to show negative refractive index. The word metamaterial was proposed by Rodger M. Wasler of University of Texas. The root of this word, "meta," is Greek, and it means beyond, i.e., metamaterial means beyond the material. Though the science of metamaterials is relatively new, Victor Veselago predicted the possibility of existence of doubly negative materials in 1968, in a Russian publication [1, 2].

Metamaterials are also known to be doubly negative materials or left-handed materials (LHM). Material with negative permeability and permittivity possess negative refractive index value, which not occurs in naturally occurring materials. Consequently, EM wave propagation through these materials follows left-handed coordinates as shown in Fig. 1.1. In a left-handed frame, the positive rotation is clockwise about the axis of rotation. The effect of this change is that it is seen that the group velocity vector and Poynting vector have opposite directions. Meanwhile, power is propagated in the opposite direction to that of the phase propagation [1]. It is also important to note that LHM is necessarily dispersive in order to meet entropy conditions. This means that either the permittivity or permeability or both vary as a function of frequency. Other properties shown by LHM include inverted Doppler effect, inverted Cerenkov effect, artificial magnetism, reversal of Snell's law.

The reversal of Snell's law can be explained through the Fig. 1.2. Consider the propagation of a ray through boundary between left-handed and right-handed

B. Choudhury (✉)
Centre for Electromagnetics, CSIR-National Aerospace Laboratories, Old Airport Road, Bengaluru 560017, India
e-mail: balamati@nal.res.in

B. Choudhury (ed.), *Metamaterial Inspired Electromagnetic Applications*,
DOI 10.1007/978-981-10-3836-5_1

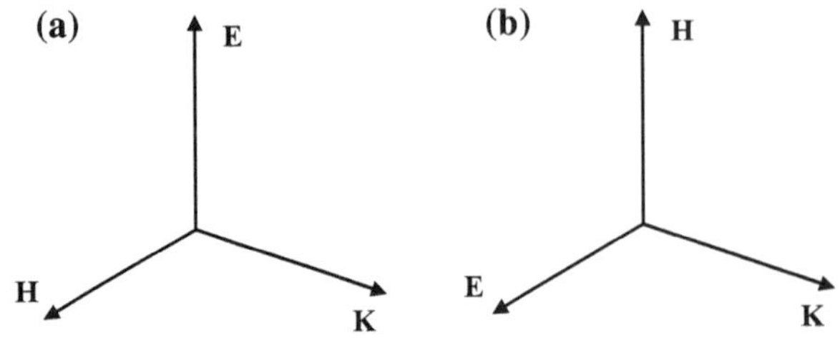

Fig. 1.1 **a** Right-handed orientation of vectors E, H, K when $\varepsilon_r > 0$, $\mu_r > 0$. **b** Left-handed orientation of vectors E, H, K when $\varepsilon_r < 0$, $\mu_r < 0$

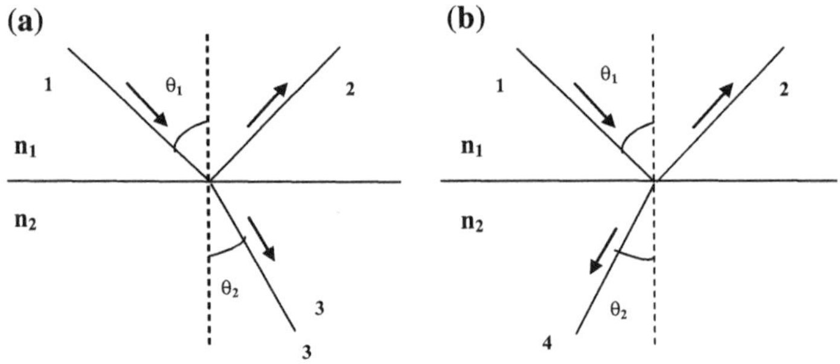

Fig. 1.2 Reflection and refraction at the interface of two media when **a** $n_2 > 0$ (ray 3), **b** $n_2 < 0$ (ray 4)

media, ray 3 is the refracted ray when the second medium is right-handed, and ray 4 is the refracted ray when the second medium is left-handed.

By considering the field components at the dielectric interfaces, if the second medium is double-negative, then the refraction of light occurs in the same side of the incident beam, i.e., behave as medium which exhibits negative effective refractive index.

The most commonly used definition of metamaterials [2] is given as "*A metamaterial is a macroscopic composite of periodic/non-periodic structure, whose function is due to both cellular architecture and chemical composition.*" The definition says that the material parameters such as permeability and permittivity of combined metamaterial structure can be engineered through the constituent materials properties and structural properties. Hence the metamaterial structures are considered as metallic patterns etched on substrate made of dielectric materials. In the literature, a number of patterns are reported, such as electric ring resonators with wires [3], graphene-based films [4], stacked rings [5], combination of FSS, I-shaped rods [6]. The dimensions of these metallic structures as well as the individual material properties and thickness of the substrate significantly affect the resonant frequency, hence resulting in a frequency-dependent material characteristics. The cellular size must be smaller than or equal to sub-wavelength. If the cell size is equal to a quarter of a wavelength, then is called effective-homogeneity limit.

It ensures that refraction dominates scattering/diffraction when wave propagates through a metamaterial.

As mentioned earlier, metamaterials are periodic/aperiodic structures composed of metallic strips etched on dielectric substrates. The geometry of the metallic layer as well as the properties of the dielectric is responsible for the resonant characteristics of the metamaterial. Figure 1.3 shows some of the commonly used metamaterial designs. Figure 1.3a is a circular split-ring resonator with a linear wire. The split-ring resonator (SRR) is responsible for the magnetic response of the metamaterial, while the linear wire in the bottom side is responsible for the electric response. Figure 1.3b shows an electric ring resonator (ELC). The resonant characteristics of the ELC are similar to those of the SRR. Figure 1.3c, d shows non-resonant metamaterials. Due to multiple rings, the metamaterial given in Fig. 1.3c resonates at three frequencies. These designs have been explored by many research groups around the world.

In order to respond to the magnetic component of the electromagnetic wave, the typically used element is split-ring resonator (SRR) and this "magnetic atom" was introduced by Pendry in 1999. According to Padilla et al. (2006), [7] in simple terms SRR can be considered as an LC resonator [7]. An EM wave with magnetic field perpendicular to plane of SRR induces circulating currents as in Faraday's law. This circulating current results charge formation across the split gap of SRR, so the energy is stored as capacitance. Hence, the SRR is considered to be as simple LC circuit with a resonant frequency. Above the resonant frequency even though external magnetic field is increased, the current can no longer keep up and gradually begin to lag, which results in a negative response or out of phase. Smith et al. (2000) [8] fabricated the negative index (NI) material by using the metamaterial

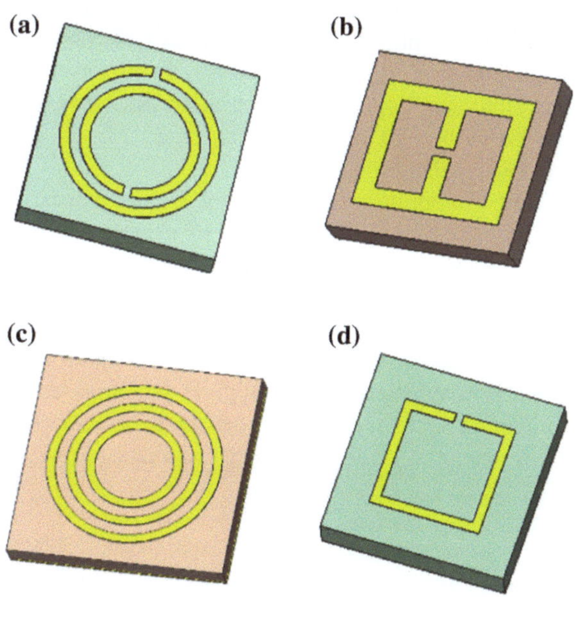

Fig. 1.3 Common metamaterial designs, **a** circular split-ring resonator, **b** electric ring resonator, **c** multi-band circular ring metamaterial structure, **d** square ring resonator

structures [8]. The structure consists of combination of a wire structure with negative permittivity and a split-ring resonator structure with negative permeability for the same frequency band, resulting in negative refraction.

Conventionally, metamaterials are simulated using software based on Finite Element Methods (FEM) or Finite Integration Techniques (FIT). This software enables the designer to study the scattering parameters (S-parameters) of the metamaterial structure and S-parameter retrieval method is used to obtain the effective material parameters. It is clear that the material characteristics are engineered through optimizing the various structural parameters such as thickness of dielectric, size, shape of the patterns in the metamaterial structure. One of the best choices for optimization in metamaterial design is the multidimensional artificial intelligent optimization algorithms.

Artificial intelligent optimization (soft computing or nature-based optimization) plays an important role in the design and optimization of various problems in engineering field. It is used to model and enable solutions to complex real-world problems for which the analytical solution methods do not exist. Soft computing is an optimization technique to find the solution of a problem which is very difficult to solve or with less mathematical formulations regarding the problem domain. The aim of these soft computing methods is to provide quick solutions which resemble human-like decisions.

The term soft computing was introduced by Zadeh (1992) [9]. Soft computing is a multidisciplinary system which includes artificial neural network (ANN), fuzzy logic (FL), and various evolutionary computing techniques such as genetic algorithm (GA), genetic programming (GP), simulated annealing (SA), particle swarm optimization (PSO), bacterial foraging optimization (BFO). Much of the soft computing techniques are inspired from biological phenomena and the social behavior of biological populations. Earlier in most literature, the term soft computing was confined to ANN, GA, and FL (Jang et al. 1997) [10]. The recently developed methods based on swarm intelligence, and foraging behavior of natural and biological populations such as birds, fishes, ants, and bacteria, are also considered to be part of the growing array of soft computing techniques ([11, 12], Passino 2002).

Soft computing methods do not require extensive mathematical formulation of the problem. Thus, the necessity of exclusive domain-specific knowledge can be reduced. Also it can handle multiple variables and multiple objective functions simultaneously. These make the soft computing techniques quite useful in EM applications to provide cost-effective solution to the user in less computational time. From last decade onwards, various soft computing techniques are used in the field of electromagnetics. This book gives an introduction to the implementation of soft computing techniques in the design of metamaterial-based applications such as PIFA antennas, cloak, and radar absorbers.

1.2 Soft Computing Techniques

Soft computing is a computational technique based on behavior of biological systems and theory of natural selection. It is used in computer science to sort the problems whose solutions are unpredictable and uncertain. The main advantages of these techniques include robustness, low-cost solution, imprecision, and uncertainty. Soft computing technique is a best choice for problems without enough information about the problem domain. Classification of optimization techniques are given in Fig. 1.4. A brief description about different soft computing techniques is given in this section.

1.2.1 Neural Network (NN)

Neural network or artificial neural network is an information processing system or a computing system which consists of highly interconnected processing elements called as neurons working in parallel to solve particular problems. The artificial neural network was first developed by McCulloch in 1943 [13]. It is inspired from the biological brain system's capability to recognize and perceive for processing information. Figure 1.5 shows a typical structure of neural network.

Fig. 1.4 Flowchart of optimization techniques

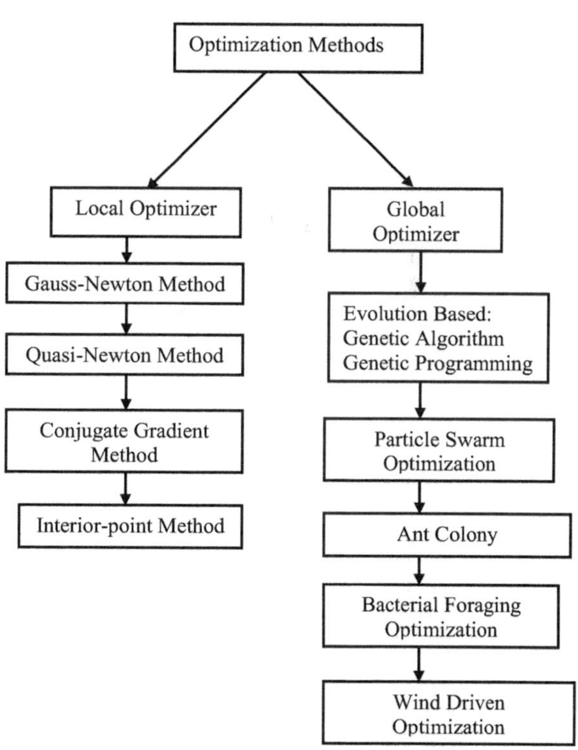

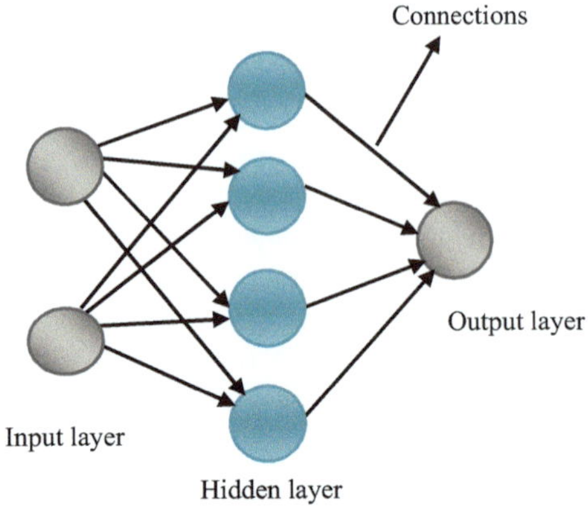

Fig. 1.5 Basic structure neural network

Neural networks are normally arranged in layers. Layers are made up of inter-connected nodes called neurons along with connection weights (strengths). The neurons are the processing nodes which contains the activation function. Patterns are presented to the network by using input layer which communicates to hidden layers, and the actual processing is done by system of weighted connections. Then, the hidden layer connects to an output layer to display the output of the network. In Fig. 1.5, each circular node represents the artificial neurons and each arrow represents a connection from the output of one neuron to input of another neuron. As like human brain intelligence, artificial neural networks gain the knowledge by learning and also the acquired knowledge from experience is stored as synaptic weights called connection strength.

The various steps in implementation of ANN are data generation, training of network and testing. In data generation phase, a large set of predetermined input output data is created and stored as biases and weights of each neuron. By using the data set available, the network learns to obtain the desired output by modifying the bias and weights. The final phase is testing of neural network by using certain test vectors. ANN can learn in both unsupervised and supervised modes. For unsupervised mode, the network self-organizes and extracts output from the dataset presented to it, whereas in supervised mode the network is provided with correct response (Wasserman 1993) [14]. ANN is best choice for solving pattern-matching problems, since it processes in parallel rather than in serial mode. The advancements in VLSI technology make the hardware implementation of these kinds of systems possible.

1.2.2 Genetic Algorithm (GA)

Genetic algorithm is an optimization procedure based on genetic inheritance and natural selection. Basic concept of optimization is based on evolution and concept of survival of the fittest.

GA is inspired by Darwin's theory of natural evolution. According to Darwin's theory, the well-suited organism survives in the natural world and poorly suited organism dies off. The algorithm starts by generating the population randomly. Population is set of initial solutions. Then, the algorithm includes three operators such as selection, crossover, and mutation.

1.2.3 Particle Swarm Optimization (PSO)

PSO is an efficient, robust, and simple optimization algorithm which is proposed by Kennedy and Eberhart in 1995 [15]. PSO is a population-based optimization algorithm based on swarm intelligence. The inspiration for the particle swarm optimization (PSO) algorithm is to find the highest density of flowers inside a field (garden) by swarm of bees. The bees start at a random location with a random velocity and optimize the path to reach their goal without any prior knowledge of the field. This behavior was modeled as an optimizer. The swarm behavior is shown below in Fig. 1.6. The selection of parameters plays an important role in optimization performance.

1.2.4 Bacterial Foraging Optimization (BFO)

Natural selection tends to eliminate animals with poor foraging strategies such as locating, handling, and ingesting food and favor the propagation of genes of those animals that have successful foraging strategies. The foraging behavior of Escherichia coli (*E. coli*) is adopted for an evolutionary computation algorithm and

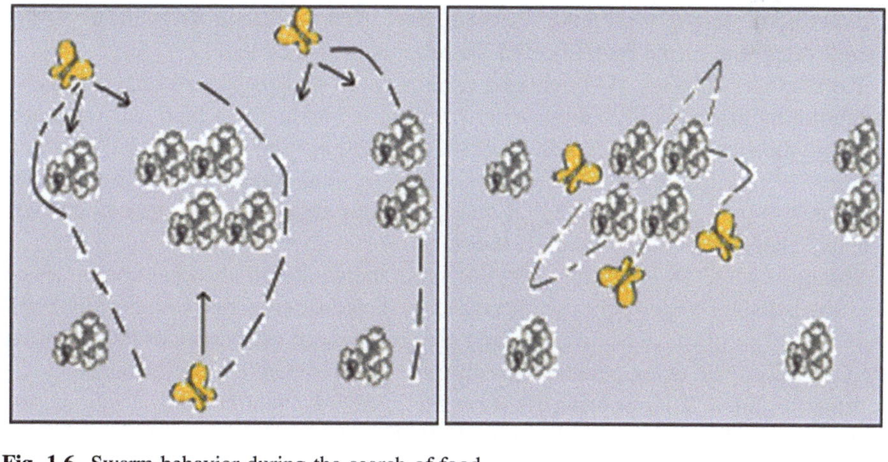

Fig. 1.6 Swarm behavior during the search of food

is named as bacterial foraging optimization (BFO). *E. coli* is a bacterium, which is having 1 μm diameter and 2 μm length, and reproduction rate of 10–7 by means of splitting in 20 min. *E. coli* has the ability to move up to 100–200 rps by means of six rigid flagella. The control system of these bacteria that dictates how foraging should proceed can be subdivided into four sections, namely chemotaxis, swarming, reproduction, and elimination and dispersal.

1.3 Trends in Soft Computing for Metamaterial Design

Soft computing methods play an important role in the design and optimization in electromagnetic applications for which analytical optimization is quite hard. In this work, soft computing techniques for metamaterial applications are carried out.

Metamaterials are artificially engineered materials made of designed structures exhibiting inimitable and exotic electromagnetic properties, which are not intrinsic in the constituent components. Metamaterial absorber is an electromagnetic absorber which absorbs all incident radiation without reflection and transmission. The performance of an absorber depends on its structural parameters such as thickness, morphology and material properties. Hence, the metamaterial absorber designs turn out to be a complex task and require optimization of various parameters. High-performance electromagnetic absorber can be implemented by exploiting the characteristics of metamaterials using various soft computing techniques. A number of literatures are available regarding metamaterial design using various soft computing techniques. This section gives a review of works done in this field.

Kern and Werner (2003a) [16] designed a metamaterial absorber to optimize high-impedance frequency selective surfaces (HZ-FSS). An ultra-thin electromagnetic band gap absorber was designed at 6 GHz. The parameters of HZ-FSS structure optimized using GA were thickness, dielectric permittivity and unit cell size. Without placing a separate resistive sheet in close proximity to the FSS, the overall thickness of the metamaterial absorber was moderated.

Kern et al. (2003c) [17] realized a thin multi-frequency artificial magnetic conductor by placing a FSS above a perfect electric conductor, which was separated by a thin dielectric layer [17]. Micro-genetic algorithm used for optimization which operates at multiple narrow frequency bands. In dual-band FSS, the operational frequency ranges were 8 and 23 GHz and in triple-band FSS the operational frequency ranges were 3.5, 11 and 18 GHz.

Huang et al. (2009) introduced an ultra-thin metamaterial absorber comprises of microstrip lines arranged periodically on a lossy dielectric substrate with metallic back plate [18]. The physical dimensions and electromagnetic properties of the substrate used are optimized using genetic algorithm for better absorption performance.

Jiang et al. (2010) presented a same kind of metamaterial absorber for single-band and dual-band applications [19]. The absorber design consisted of two metallic layers separated by a thin dielectric spacer. The top layer has an array of patterned metallic screens, and a bottom layer has continuous metallic sheets. The geometrical parameters of the metamaterial absorber were optimized using genetic algorithm to provide an absorption greater than 0.94.

Guodos et al. (2006) used multi-objective particle swarm optimization (MOPSO) to design a planar multilayer coating with high absorption for desired range of frequencies and angles [20]. Multi-objective PSO (MOPSO) was more efficient and faster compared to multi-objective genetic algorithm.

Choudhury et al. (2013) used particle swarm optimization for designing metamaterial absorber for terahertz frequency applications [21]. The optimized design results enhanced absorption characteristics; around 99.32% absorption is obtained at the required frequency of operation.

Bacterial foraging algorithm is still new to applications in the field of metamaterial technology. Choudhury et al. (2013) use BFO for design and optimization of circular split-ring resonator [22]. By using BFO, the structural parameters of CSRR are estimated at the required operating frequency and better absorption is obtained. Also a comparison study of computational time and accuracy of various soft computing techniques are discussed.

1.4 Design of Metamaterial Structure Incorporating Soft Computing Optimization Techniques

This section deals with the design of various metamaterial structures along with soft computing based optimization techniques.

1.4.1 Genetic Algorithm

Genetic algorithm is regarded as a general purpose search algorithm and as a robust optimization procedure works based on process of natural selection. In 1975, John Holland developed this algorithm inspired from Darwin's theory about natural evolution, the concept of survival of fittest among population [23]. The goal of GA is to understand the adaptive processes of natural systems and to design artificial systems software that holds the robustness of natural systems. Genetic algorithm is used for solving complex problems for which analytical solution methods are not feasible. The implementation of genetic algorithm begins with a population of chromosomes. The goodness of a solution is described with respect to the current population.

Genetic algorithms differ from other ordinary search algorithms in the following way:

- Instead of assigning exact values to the parameter, coding of parameters is used (real or binary).
- Searching is done through population of points rather than through single points.

- Instead of using derivatives, GA uses evolution function called fitness function.
- The rules used in GA are probabilistic in nature rather than deterministic.
- Probabilistic rules are used instead of the deterministic ones.

The most important parameters in genetic algorithm are population size, evolution function (fitness function), crossover and the mutation rate. The basic workflow of genetic algorithm is shown in Fig. 1.7.

- **Population**:

Population is defined as collection of individuals. Each state of population is called as generation that undergoes the changes to give new generation. The basic concept of GA is to maintain a population of chromosomes with a static population size. A chromosome is a collection of genes. Each individual chromosome gives possible solution to the optimization problem. Different methods are used to represent the chromosomes such as binary encoding, value encoding, permutation encoding. Usually, binary encoding is used in which individual values are encoded as bit string consist of binary values either 0 or 1. The dimension of GA is referred by dimension of search space which is equal to the number of genes in each chromosome.

- **Fitness**:

Fitness function takes main aspect in every optimization problem. The fitness function provides the way to evaluate the performance of each chromosome in the population, that is, the ability of chromosomes to compete in the population. Fitness

Fig. 1.7 Flow of genetic algorithm

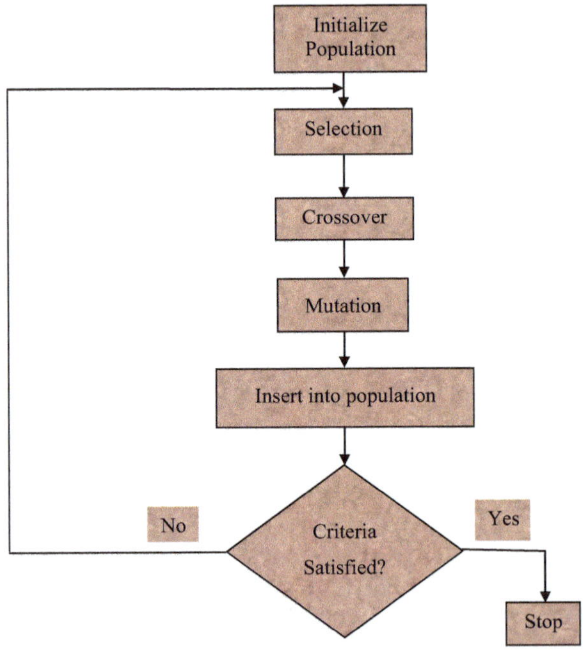

evaluation function is either complex or simple, and it depends on optimization problem. For a particular chromosome, the fitness function returns a single value which denotes the merit of corresponding solution to the given problem. As the algorithm maintains static population size, the chromosomes with low fitness value are eliminated and give space for offspring with better fitness.

- **Selection**:

Selection process is used to select the chromosomes from the current generation to be the parents for next generation, so that the offspring generated have better fitness than their parents. Different methods used to select the parent chromosomes are uniform selection, Roulette wheel selection and tournament selection, etc. Out of these methods, most commonly used selection method is Roulette wheel selection.

– Roulette wheel selection:

In this selection method, the individual which has greatest fitness value has the highest chances of selection. Each individual is allocated in the spaces on a Roulette wheel according to their fitness value.

Algorithm of Roulette wheel selection is given below:

(i) Do summation of fitness for all members of population and store the results in *S*.
(ii) Generate a random number *R* between 0 and *S*.
(iii) Go through the population and sum the fitness from *0* to *S*. Check whether *S* is greater than *R*, then stop, and return to the corresponding chromosome.
(iv) Repeat the steps from (ii) to (iii) until the number of chromosomes and size of population become equal.

So here the total fitness score of the population is represented by a Roulette wheel. A part or slice of a wheel is allocated to each chromosome of the population. The size of the slice is proportional to that member or chromosomes fitness value. Then the chromosome number can be identified though the position of the spin ball that stopped after rotation of the roulett wheel.

- **Crossover**:

Crossover is a genetic operator or a process of taking more than one parent solutions to produce a offspring solution from them. Along with selecting the parent chromosomes, the crossover site is also selected randomly along the string. After the crossover points, the genes of parents are exchanged, so that new offspring is created and parents are eliminated. A number of crossover techniques are there, which are given below.

– Single point crossover:

In this crossover, a single crossover point on both parents is selected randomly and is illustrated in Fig. 1.8. Data from the starting of the chromosome to the crossover point are copied from the first parent, and the rest are copied from the other parent.

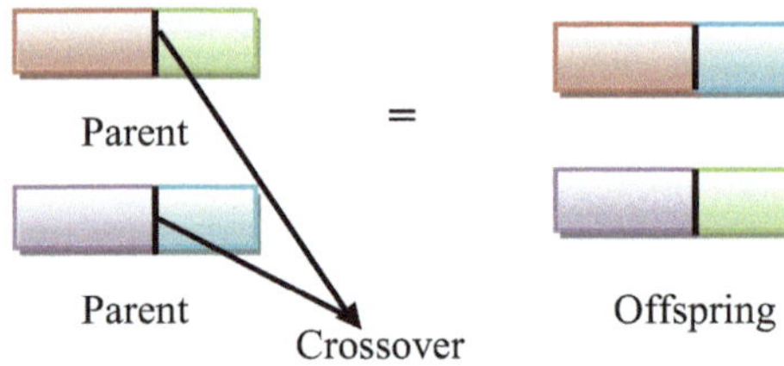

Fig. 1.8 Single point crossover

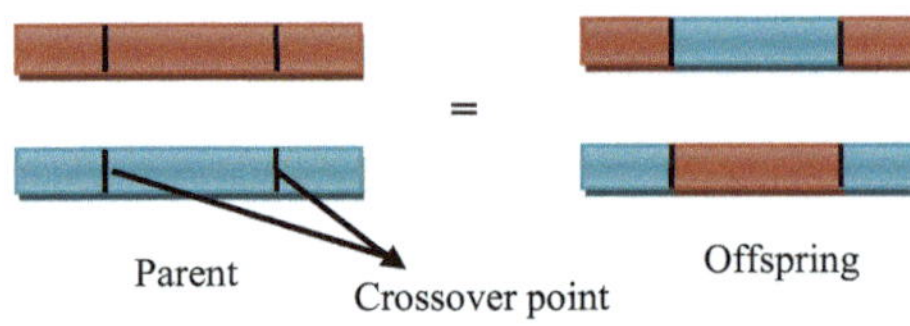

Fig. 1.9 Two-point crossover

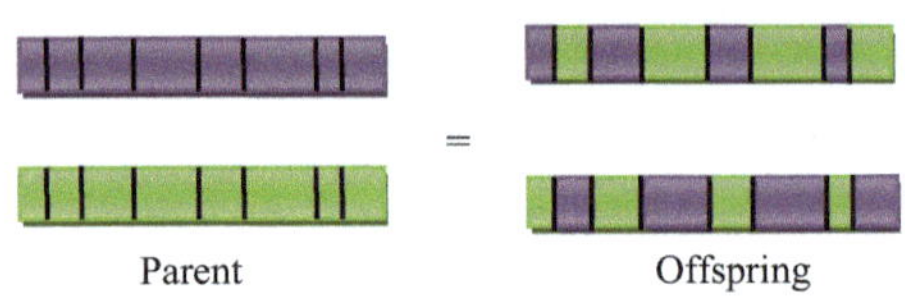

Fig. 1.10 Uniform crossover

– Two-point crossover:

Crossover operator randomly selects two crossover points in the chromosome. Binary string will be selected from the first parent from the beginning of chromosome to the first crossover point. Then the binary string from first crossover point to second crossover point will be taken from the second parent, and rest will be taken from the first parent. The schematic representation is shown in Fig. 1.9.

– Uniform Crossover:

In uniform crossover, bits are selected randomly from the first or second parent as shown in Fig. 1.10. The amount of crossover is controlled by crossover probability. If the mixing ratio is 0.5, then the offspring will contain the half gene from the first parent and other half from the second parent.

- **Mutation**:

Mutation is a genetic operator which selects a random position of a random chromosome and replaces the corresponding gene or bit by some other information or through flipping of bits as shown in Fig. 1.11. It requires only one parent to generate the offspring. Mutation results either offspring with desirable traits or offspring that are not suitable for environment as compared to the parent chromosomes. Through natural selection, the former propagate to the next generation

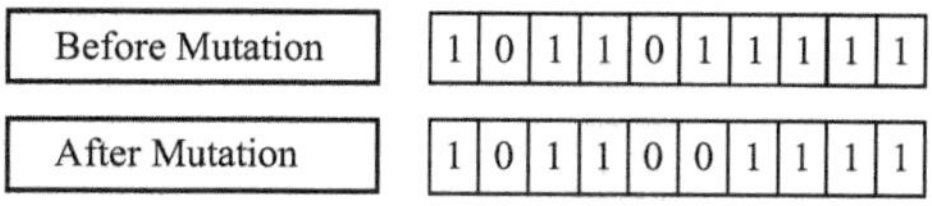

Fig. 1.11 Mutation operation in genetic algorithm

and exterminated the latter. Probability of mutation is determined before starting of the algorithm which is applied to each individual bit of each offspring chromosome to determine whether it is to be inverted.

The rate at which the new gene values propagate to next generation is controlled through mutation probability. Premature convergence such as solutions fall into local optimum value is avoided by using mutation along with crossover operation.

1.4.2 Particle Swarm Optimization (PSO)

PSO is an efficient, robust, and simple optimization algorithm which is proposed by Kennedy and Eberhart in 1995 [15]. The inspiration for the particle swarm optimization (PSO) algorithm is to find the highest density of flowers inside a field (garden) by swarm of bees. The bees start at a random location with a random velocity and optimize the path to reach their goal without any prior knowledge of the field. This behavior was modeled as an optimizer. The selection of parameters plays an important role in optimization performance.

In PSO, the key terms used are given below

- *Swarm*: It is the entire collection of bees (particles) in the field. It relates to the search space for the particular optimization problem.
- *Particle or Agent*: The individual bees in the swarm are referred as particle. All the particles in the swarm act individually and accelerate toward the best overall location.
- *Position or Location*: This refers to the particle's spatial place within the field. A particle's N-dimensional coordinates represent a solution to the problem.
- *Fitness*: Fitness is defined as a single number representing the accuracy of a given solution. The fitness function must take the position in the solution space and return a single number representing the value of that position. This function provides the interface between the physical problem and optimization algorithm.
- p_{best} *(personal best)*: It is defined as the best position found by the particle. Each particle has its own personal best. It is updated whenever the particle reaches a position with better fitness value than the fitness value of the previous personal best.
- g_{best} *(global best)*: It is defined as the best position found by all the particles in the swarm. For entire swarm, there is only one global best. It will be updated whenever a particle reaches a position with better fitness value than the fitness value of the previous global best.

Algorithm

The algorithm of PSO starts with defining solution space. Initially, the number of parameters to be optimized and fitness function are defined along with various constants involved in the algorithm. The algorithm steps are given below.

Step 1: Define the solution space. The parameters to be optimized are selected in this space. A minimum ($x_{min}(n)$) and a maximum ($x_{max}(n)$) range is defined, where n ranges from 1 to N (dimension of the optimization space).

$$v_{\min} = -0.2(x_{\max} - x_{\min}) \tag{1.1}$$

$$v_{\max} = 0.2(x_{\max} - x_{\min}) \tag{1.2}$$

Step 2: Define a fitness function: The fitness function exhibits a functional dependence that is relative to the importance of each characteristic being optimized.

Step 3: Initialize random swarm location and velocities. Each particle begins at its own random location with a velocity that is random in its direction and magnitude. Also p_{best} and g_{best} are found.

Step 4: Systematically fly the particles through the solution space. The algorithm acts on each particle one by one, moving it by a small amount and cycling through the entire swarm. The following steps are encountered on each particle.

- Evaluate the particle's fitness and compare to g_{best} and p_{best}.
- Update the particle's velocity (v) using the following equation.

$$v = w \times v + c_1 r_1 (p_{\text{best}} - x) + c_2 r_2 (g_{\text{best}} - x) \tag{1.3}$$

where c_1 is called the social constant and c_2 is called cognitive constant and w is the inertial constant. From these equations, it is understood that movement of particles is explained using Newtonian mechanics. For better results, the inertial constant value w is decremented from 1 to 0.4 during the start of algorithm to end. Also the values of social and cognitive constants are set to 0.5 to increase the convergence rate of algorithm.

- Move the particle to next location: Once the particle velocity has been updated, the particle has to move to its next location. The velocity is applied for time-step t, and new coordinate x is computed for each of the N dimensions according to the following equation.

$$x = x + v \tag{1.4}$$

Step 5: For each particle in the swarm, step 4 is repeated. Every second, the snapshot is taken for the entire swarm, such that the time and the positions of all particles are evaluated and modifications are made to p_{best} and g_{best} values if required.

The flowchart of the PSO algorithm is shown in Fig. 1.12.

Binary, Continuous, and Hybrid Optimizations:
According to the optimization variables, the PSO is classified to be binary, continuous and hybrid optimizations. A set of real values for optimization variables are determined from the infinite solution domain for continuous PSO algorithm.

In each application, the PSO algorithm is used for optimization of different parameters for different fitness function or objective function. Sometimes more than one variable or multiple objective functions need to be optimized.

Application:
Analysis of metamaterial structures is carried out using equivalent circuit analysis method. Here, double-ring square SRR is analyzed. PSO is used to optimize the size of double-ring square SRR in order to find out minimum error in resonant frequency. The parameters that are optimized are the three dimensions of square SRR such as side length, conductor width and dielectric spacing between inner and outer ring. Optimization is done with respect to a desired resonant frequency.

Double-ring square SRR:
A double-ring square SRR is a metamaterial structure, which consists of two square-shaped rings with the gaps in between them on the opposite sides of both rings. Figure 1.13 shows the structure of double-ring square SRR, which is generally printed on a dielectric substrate. Here a, w, d denote the side length of the square, width of conductor, space between the inner and outer squares, respectively, and g is the gap present in the rings. The gap width does not vary from the inner ring to the outer ring.

Equivalent Circuit Analysis:
The square SRR can be represented as a parallel LC tank circuit. The resonant frequency of a square SRR is obtained using equivalent circuit analysis method [7]. In this method, the distributed network is converted to lumped network as shown in Fig. 1.14 and the analysis has been carried out.

From the analysis of the above equivalent circuit, the resonant frequency of the double-ring square SRR is given by

$$f_0 = \frac{1}{2\pi\sqrt{L C_s}} \tag{1.5}$$

where

C_s is the equivalent capacitance
L is the effective inductance due to both rings.

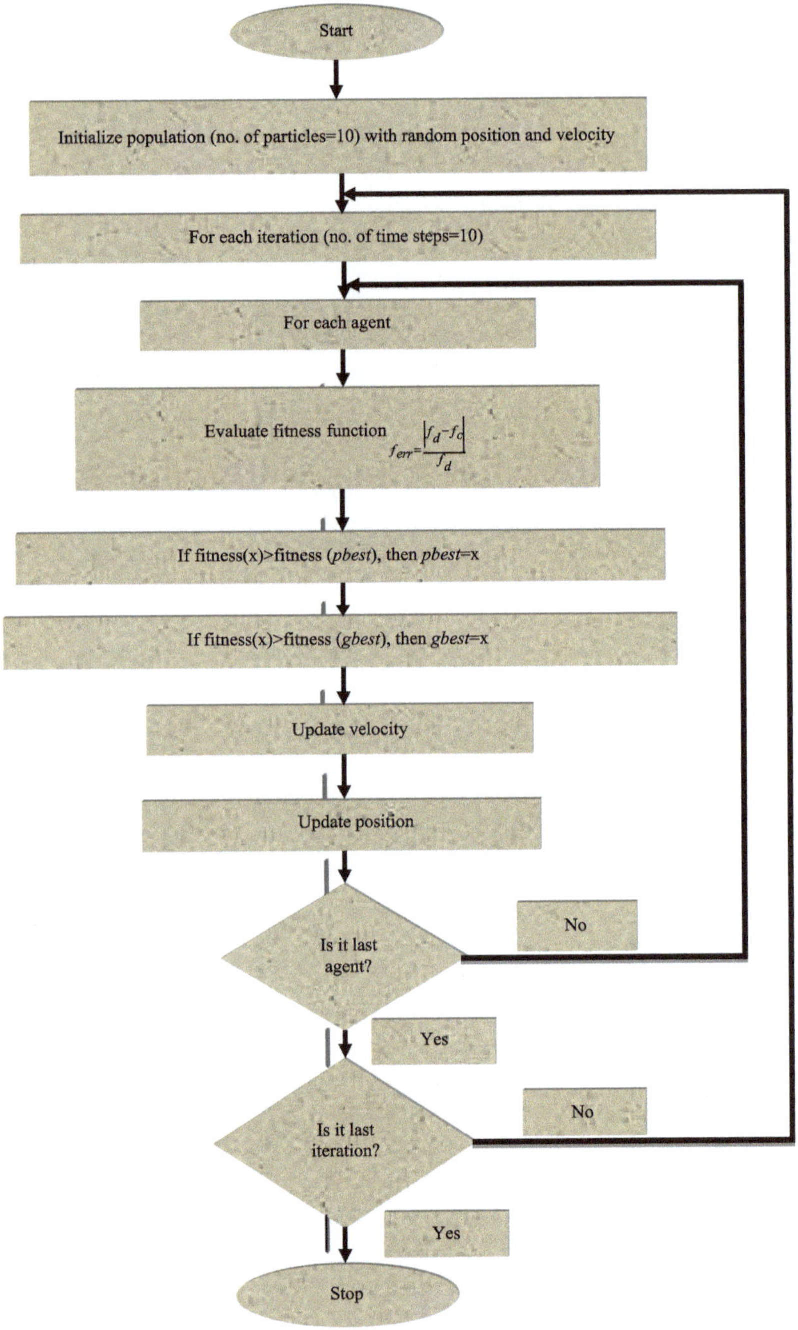

Fig. 1.12 Flowchart of PSO algorithm

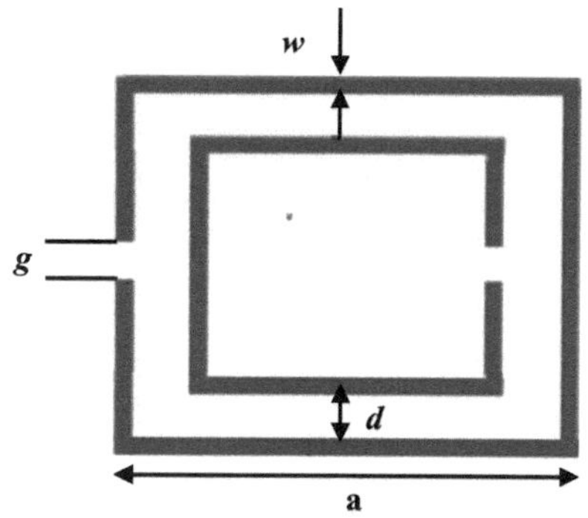

Fig. 1.13 Structure of double square split-ring resonator

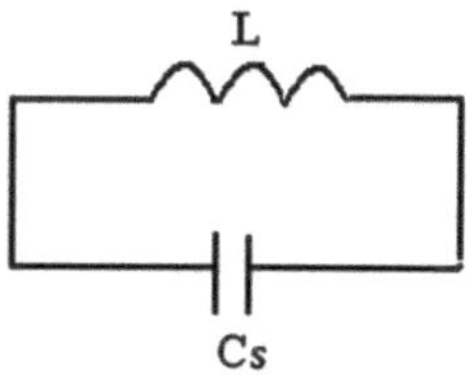

Fig. 1.14 Equivalent circuit of square split-ring resonator

The effective inductance of the square SRR is given by [8]

$$L = \frac{4.86\,\mu_0}{2}(a - w - d)\left[\ln\left(\frac{0.98}{\rho}\right) + 1.84\rho\right] \tag{1.6}$$

ρ is the filling factor of the inductance and is given by

$$\rho = \frac{w + d}{a - w - d} \tag{1.7}$$

The effective capacitance is given by

$$C_s = \left(a - \frac{3}{2}(w + d)\right) C_{\text{pul}} \tag{1.8}$$

where C_{pul} is the per-unit-length capacitance between the rings which is given as below

$$C_{\text{pul}} = \varepsilon_0\,\varepsilon_{\text{eff}} \frac{K\left(\sqrt{1 - k^2}\right)}{K(k)} \tag{1.9}$$

Here, ε_{eff} is the effective dielectric constant which is expressed as

$$\varepsilon_{\text{eff}} = \frac{\varepsilon_r + 1}{2} \tag{1.10}$$

$K(k)$ denotes the complete elliptical integral of the first kind with k expressed as

$$k = \frac{d}{d + 2w} \tag{1.11}$$

The accuracy and efficiency of PSO algorithm optimization depend on cost function formulation. The cost function to be optimized is

$$f_{\text{er}} = \frac{|f_{\text{desired}} - f_{\text{calc}}|}{f_{\text{desired}}} \tag{1.12}$$

where f_{calc} denotes the obtained calculated frequency through equivalent circuit analysis and $f_{desired}$ is the desired frequency.

The algorithm steps implemented for this cost function optimization and corresponding structural parameters such as square SRR, side length of the square, width of conductor, space between the inner and outer squares and g the gap present in the rings is obtained. These parameters are given as the input for simulation studies, and output responses are obtained.

Simulation and Validation of double-ring square SRR

Validation example of double-ring square SRR is described through simulation studies having resonant frequency 9.6 GHz and dielectric constant of the substrate as 4.4. The unit cell structure is as shown in Fig. 1.15. The simulation is carried out for a side length a = 2.8 mm, width of the ring w = 0.3 mm, and gap in the rings g = 0.3 mm. After obtaining the structural parameters, the simulation studies were carried out and the corresponding permittivity and permeability have been extracted.

The simulation results are shown in Fig. 1.16a, b.

It is clear that the designed metamaterial square double-ring SRR has negative permeability from 9.35 to 9.94 GHz, which fulfills the design objective.

Fig. 1.15 PSO optimized double-ring square SRR designed using FEM solver

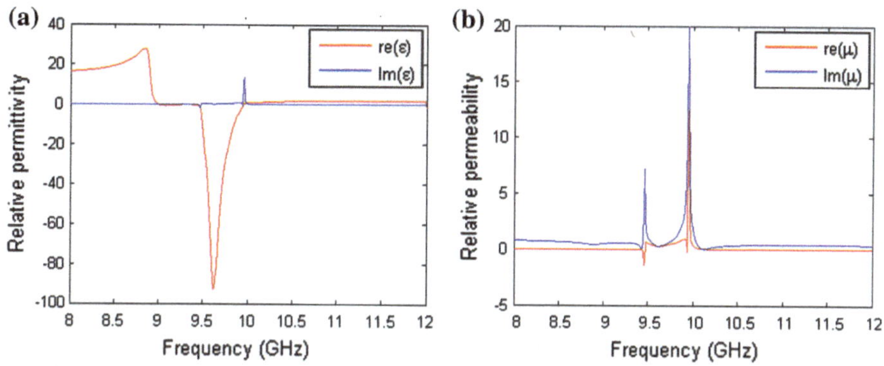

Fig. 1.16 Extracted characteristics of the modeled double-ring square SRR, **a** permittivity, **b** permeability

1.4.3 Bacterial Foraging Optimization (BFO)

As compared to GA and PSO, the bacterial foraging optimization (BFO) is fairly new in the field of applications in electromagnetics. K.M. Passino introduced this technique in 2002 [12]. The algorithm is inspired from the imitation of food-ingesting (foraging) *E-coli* bacteria, which are present in our intestines. Here, a group of bacteria moves in search of rich nutrient concentration and away from noxious elements. The BFO proceeds by selecting or eliminating bacteria based on their foraging strategies. The natural selection tends to eliminate animals with poor foraging strategies and favor those having the successful ones. After many generations, the poor foraging strategies are either eliminated altogether or reshaped into the good ones. The foraging strategy is governed by four different steps which includes chemotaxis, swarming, reproduction, and elimination–dispersal are given below.

- **Chemotaxis**:

This process is achieved through swimming and tumbling by flagella. Depending upon the rotation of flagella in each bacterium, it can move in two different ways; it can move in a predefined direction (swimming) or altogether in different directions (tumbling), in the entire lifetime. To represent a tumble, a unit length random direction, say $\varphi(j)$, is generated; this will be used to define the direction of movement after a tumble.

In particular

$$\theta^i(j+1,k,l) = \theta^i(j,k,l) + C(i)\phi(j) \qquad (1.13)$$

where $\theta^i(j, k, l)$ represents the ith bacterium at j^{th} chemotactic, k^{th} reproductive and l^{th} elimination and dispersal step. $C(i)$ is the size of the step taken in the random direction specified by the tumble (run length unit).

- **Swarming**:

During the process of reaching toward the best food location, it is always desired that the bacterium which has searched the optimum path should try to provide an attraction signal to other bacteria so that they swarm together to reach the desired location. In this process, the bacteria congregate into groups with high bacterial density. The mathematical representation for swarming can be represented by

$$\begin{aligned} J_{cc}(\theta, P(j,k,l)) &= \sum_{i=1}^{S} J_{cc}^{i}(\theta, \theta^{i}(j,k,l)) \\ &= \sum_{i=1}^{S}\left[-d_{\text{attract}}\exp\left(-\omega_{\text{attract}}\sum_{m=1}^{p}\left(\theta_m-\theta_m^i\right)^2\right)\right] \\ &\quad+\sum_{i=1}^{S}\left[h_{\text{repellent}}\exp\left(-\omega_{\text{repellent}}\sum_{m=1}^{p}\left(\theta_m-\theta_m^i\right)^2\right)\right] \end{aligned} \quad (1.14)$$

where J_{cc} (θ, P (j, k, l)) is the cost function value added to the actual cost function which is to be minimized to present a time varying cost function. "S" is the total number of bacteria, and "p" is the number of parameters to be optimized which are present in each bacterium. $\omega_{attract}$, $d_{attract}$, $h_{repellent}$, and $\omega_{repellent}$ are different coefficients that are to be chosen properly.

- **Reproduction**:

The least healthy bacteria die, and the other healthiest bacteria will split into two bacteria, which are placed in the same location. This makes the population of bacteria constant.

- **Elimination and Dispersal**:

The two processes which determine the population of bacteria are elimination and dispersal. Gradually, the elimination occurs due to consumption of nutrients in particular area and dispersal results in the spreading of bacteria from one area to another area. These two processes destroy the progress of chemotactic step. This process made the BFO prevent the algorithm from local optima by preventing stagnation.

Algorithm

The algorithm of BFO is as follows

Step 1: Initialize parameters p, S, N_c, N_s, N_{re}, N_{ed}, P_{ed}, $C(i)$ (i = 1, 2,…S), θ^i where,

p	Dimension of search space
S	Total number of bacteria in the population
N_c	The number of chemotaxis steps
N_s	The number of swimming steps

N_{re} The number of reproduction steps
N_{ed} The Number of elimination and dispersal steps
P_{ed} Elimination–dispersal with probability
$C(i)$ The size of the step taken in the random direction specified by the tumble

Step 2: Elimination–dispersal loop: $l = l + 1$
Step 3: Reproduction loop: $k = k + 1$
Step 4: Chemotaxis loop: $j = j + 1$

(i) For $i = 1, 2, \ldots S$, take a chemotactic step for bacterium i as follows.
(ii) Compute fitness function, $J(i, j, k, l)$.
Let, $J(i, j, k, l) = J(i, j, k, l) + J_{cc}(\theta^{i}(j, k, l), P(j, k, l))$ (i.e., add on the cell-to-cell attractant-repellant profile to simulate the swarming behavior)
where J_{cc} is defined in (ii).
(iii) Let $J_{last} = J(i, j, k, l)$ to save this value since we may find a better cost via a run.
(iv) Tumble: Generate a random vector $\Delta(i) \in \Re^{p}$ with each element $\Delta_m(i)$, $m = 1, 2, \ldots, p$, a random number on $[-1, 1]$
(v) Move: Let

$$\theta^{i}(j+1, k, l) = \theta^{i}(j, k, l) + C(i)\frac{\Delta(i)}{\sqrt{\Delta^{T}(i)\Delta(i)}} \quad (1.15)$$

This results in a step size $C(i)$ in the direction of the tumble for bacterium i.
(vi) Compute $J(i, j + 1, k, l)$ and let

$$J(i,j+1,k,l) = J(i,j,k,l) + J_{cc}(\theta^{i}(j+1,k,l), P(j+1,k,l)) \quad (1.16)$$

(vii) Swim

(a) Let $m = 0$ (counter for swim length)
(b) While $m < N_s$ (if have not climbed down too long)

- Let $m = m + 1$
- If $J(i, j + 1, k, l) < J_{last}$ (if doing better), let $J_{last} = J(i, j + 1, k, l)$ and let

$$\theta^{i}(j+1, k, l) = \theta^{i}(j, k, l) + C(i)\frac{\Delta(i)}{\sqrt{\Delta^{T}(i)\Delta(i)}} \quad (1.17)$$

- Else, let $m = N_s$. This is the end of the while statement

(viii) Go to the next bacterium $(i + 1)$ if $i \neq S$ (i.e., go to (ii)) to process the next bacterium.

Step 5: If $j < N_c$, go to Step 4. In this case, continue chemotaxis since the life of the bacteria is not over.

Step 6: Reproduction:

(i) for the given k and l, and for each $i = 1, 2, \ldots S$, let

$$J^i_{\text{health}} = \sum_{j=1}^{N_c+1} J(i,j,k,l) \tag{1.18}$$

Be the health of the bacterium i (a measure of how many nutrients it got over its lifetime and how successful it was at avoiding noxious substances). Sort bacteria and chemotactic parameters $C(i)$ in the order of ascending cost J_{health} (higher cost means lower health).

(ii) The S_r bacteria with the highest J_{health} values die and remaining S_r bacteria with the best values split (this process is performed by the copies that are placed at the same location as their parent).

Step 7: If $k < N_{re}$, go to Step 3. In this case, we have not reached the number of specified reproduction steps, so we start the next generation of the chemotactic loop.

Step 8: Elimination–dispersal: For $i = 1, 2\ldots, S$ with probability P_{ed}, eliminate and disperse each bacterium (this keeps the number of bacteria in the population constant). To do this, if a bacterium is eliminated, simply disperse another one to a random location on the optimization domain.

Step 9: If $l < N_{ed}$, then go to Step 2; otherwise end.

Application:
In metamaterial design, the basic structure used is split-ring resonators which possess negative refractive index at the corresponding frequency of resonance. Analysis of metamaterial structures has been carried out using equivalent circuit analysis method. The structural parameters such as diameter, width, gap between rings and material properties of the substrate determine the frequency of resonance. Here, double-ring circular SRR is analyzed. Bacterial foraging optimization (BFO) algorithm is implemented for design optimization of circular SRR with respect to a desired resonant frequency.

Circular SRR:
Figure 1.17a represents the schematic of the circular SRR along with dimensions. The external radius is represented by r_{ext}, width of rings is denoted by w, d is the separation between inner and outer rings, and s is the gap in each ring. For the

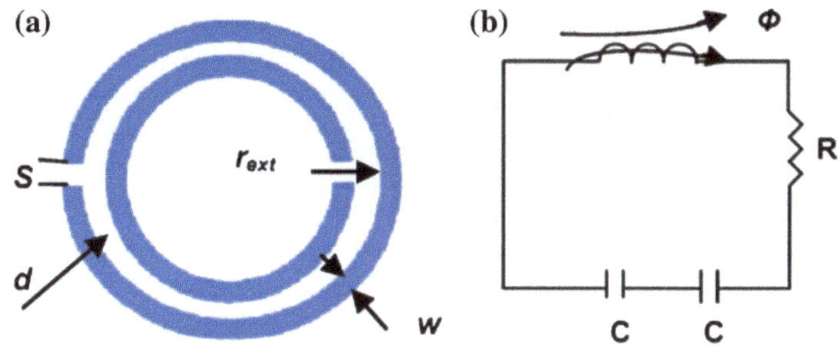

Fig. 1.17 **a** Schematic of circular SRR along with dimensions. **b** Distributed circuit represented by corresponding lumped network

analysis, the distributed circuit is represented by the corresponding lumped network and is shown in Fig. 1.17b.

From the equivalent circuit of circular SRR, the resonance frequency of operation is

$$f_r = \frac{1}{2\pi\sqrt{LC}} \quad (1.19)$$

where C and L are the capacitance and inductance of the CSRR, respectively.

Formulation of problem:

The accuracy and efficiency of BFO algorithm implementation depend on cost function formulation. The cost function to be optimized is

$$f_{\text{er}} = \frac{|f_{\text{desired}} - f_{\text{calc}}|}{f_{\text{desired}}} \quad (1.20)$$

where f_{calc} denotes the obtained calculated frequency through equivalent circuit analysis and $f_{desired}$ is the desired frequency.

The algorithm steps implemented for this cost function optimization and corresponding structural parameters such as CSRR external radius, width of rings and separation between rings are obtained. These parameters are given as the input for simulation studies, and output responses are obtained.

Also a feasibility study is carried out among GA, PSO, and BFO; results show that in terms of CPU time and accuracy, BFO is more efficient. Table 1.1 shows the comparison between various soft computing techniques. The CPU time and accuracy in any soft computing technique depend on the selection of parameters, complexity and type of problem.

Table 1.1 Comparison of soft computing techniques for optimization of a CSRR at 8.25 GHz (Intel core, 4 GB RAM)

Techniques	Accuracy (f_{error} in GHz)	CPU time (in s)
BFO	0.0181	0.058939
PSO	0.0285	0.213547
Genetic algorithm	1.5710	0.226509

By using the obtained parameters from optimization through BFO algorithm, simulation studies are carried out. From the obtained results, permeability and permittivity are also extracted for ensuring the metamaterial properties.

Simulation Results

The metamaterial dual circular SRR is designed for a resonant frequency of 4.5 GHz. The dielectric constant of substrate used is 3.8, and other structural parameters are height of substrate of 2.4 mm, external radius of 2.7 mm, width of ring w of 0.3 mm, and gap in each ring as 0.3 mm. Figure 1.18 shows the unit cell of the CSRR designed. The obtained S-parameter shows that resonance frequency is obtained at 4.5 GHz and fulfills the objective of design. The reflection and transmission characteristics of the designed structure are shown in Fig. 1.19.

The extracted material parameters are shown below in Figs. 1.20 and 1.21. From the responses, it is clear that permeability of the metamaterial is negative in the range of frequency of resonance (Fig. 1.21).

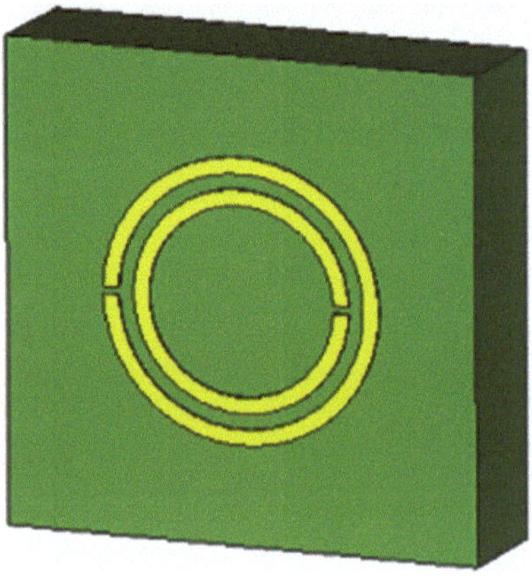

Fig. 1.18 Unit cell of the CSRR designed

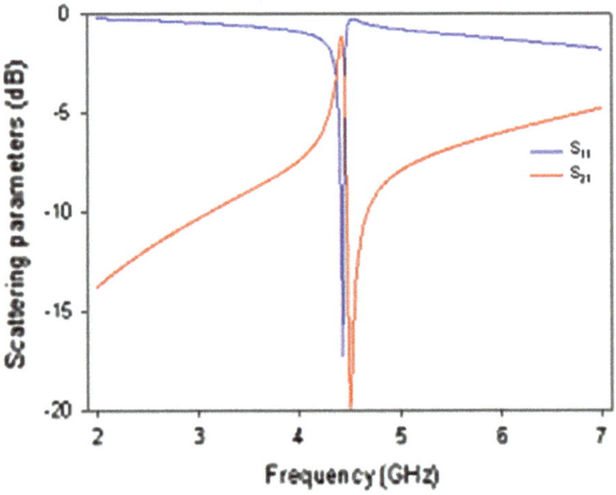

Fig. 1.19 Reflection and transmission characteristics of the designed CSRR structure

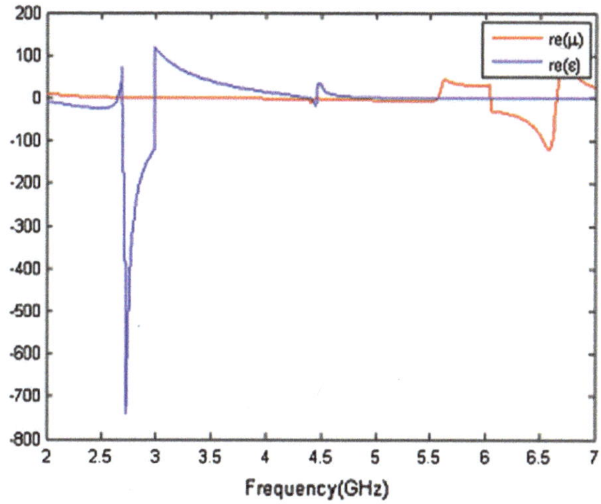

Fig. 1.20 Extracted permittivity and permeability of the designed CSRR

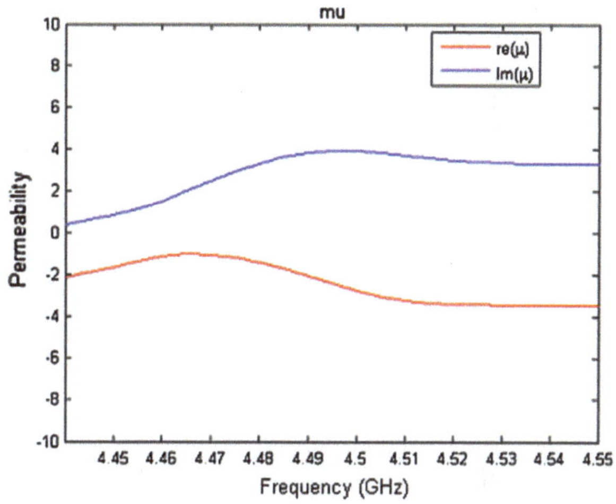

Fig. 1.21 Negative permeability at the desired resonant frequency

1.5 Linking Optimization Algorithm to EM Tools

The lack of mathematical formulation in EM domain leads to the integration of the optimization tool with EM simulators in order to achieve optimization through iterative EM simulations. In this work, the soft computing optimization code acts as a kernel, which calls a FEM-based EM solver, in order to determine the fitness as shown in the Fig. 1.22. This technique involves calling the solver-specific commands (such as visual basic language (VBA) EM simulation solver) in the Matlab framework. The soft computing optimization algorithm, along with the EM simulation tool, creates a computational engine for optimization.

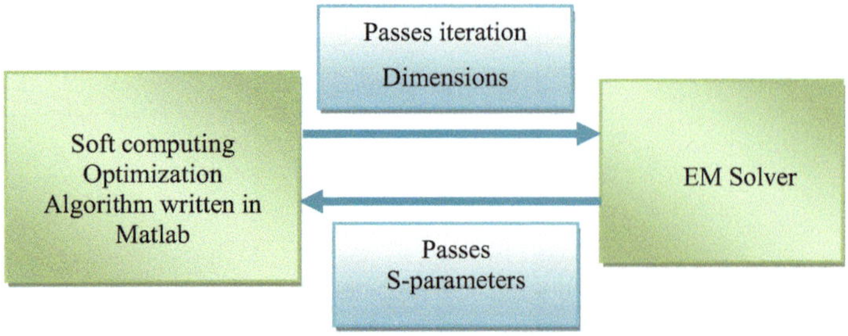

Fig. 1.22 Schematic of computational engine based on optimization algorithm

The developed computation engine passes the coordinates of each particle (structural parameters) during every step in the iteration to EM solver. Using this dimensional information, and solver-specific commands, which involve assigning of material properties to each component in the structure, the solver is able to design the structure, assign boundary conditions, assign EM solver parameters, simulate the structure for a particular frequency range and output the S-parameters obtained from the simulation. These S-parameters are sent back to the algorithm kernel, where these values are used to calculate an application-specific fitness function.

1.6 Conclusion

The basic concepts of soft computing algorithms have been reported in this chapter in detail along with their stepwise development procedure in terms of flowchart. The same algorithms have been implemented in design and analysis of metamaterials in conjunction with equivalent circuit model as electromagnetic tool. Nature-based algorithms being the best intelligent algorithms, genetic algorithm, particle swarm optimization and bacterial foraging optimization techniques are considered in this book. The algorithms described in this chapter will be the basis for the subsequent chapter where real-time electromagnetic design optimization applications have been demonstrated.

References

1. Veselago, V.G., "The electrodynamics of substances with simultaneously negative values of e and m," *Soviet Physics Uspekhi*, vol. 10, no. 4, pp. 509–514, Jan.–Feb. 1968.
2. Cui,T. J., D.R. Smith, and R. Liu, *Metamaterial Theory and Design*, Springer, NewYork, ISBN: 978-1-4419-0573-4, 367p., 2010.
3. Yang, T., X. Li, and W. Zhu, "A tunablemetamaterial absorber employing MEMSactuators in THz regime," *NEMS*, pp. 829–832, China, Apr. 2013
4. He, S., "Broadband THz absorbers with graphene-based anisotropic metamaterialfilms," *IEEE Transactions on Terahertz Science and Technology*, 7 p., 2013.

5. Kung, P., and S. M. Kim, "Terahertz metamaterial absorbers for sensing and imaging," *Proceedings of Progress in Electromagnetics Research Symposium*, pp.232–235, Mar. 2013.
6. Hokmabadi, M. P., D. S. Wilbert, P. Kung, and S. M. Kim, "Terahertz metamaterialabsorbers," *Terahertz Science and Technology*, vol. 6, no. 1, pp. 40–58, Mar. 2013.
7. Padilla, W. J., A. J. Taylor, C. Highstrete, M. Lee, and R. D. Averitt, "Dynamical electric metamaterial response at terahertz frequencies," *Proceedings of the 15th International Conference*, Pacific Grove, USA, pp. 642–644, Aug. 2006.
8. Smith, D. R., J. B. Pendry, and M. C. K. Wiltshire, "Metamaterials and negative refractive index," *Science*, vol. 305, pp. 788–792, Aug. 2000.
9. Zadeh, L. A., "Fuzzy logic, neural networks and soft computing," *One-page course announcement of CS 294-4, Spring 1993*, The university of California at Berkely, November 1992.
10. Jang, J. S. R., C. T. Sun, and E. Mizutani, *Neuro-Fuzzy and Soft Computing*, Prentice Hall, NJ, ISBN:9780132610667, 614p., 1997.
11. Passino, K. M., "Biomimicry of bacteria foraging for distributed optimization and control," *IEEE Control system Magazine*, vol.22 pp. 52–67, 2002.
12. Passino, K. M., *Biomimicry for Optimization, Control and Automation*, Springer-Verlag, London, ISBN: 1-85233-804-0, 926p., 2005.
13. McCulloch, W. S., and W. Pitts, "A logical calculus of the ideas immanent in nervous activity," *Bulletin of Mathematical Biophysics*, vol. 5, pp. 115–133, 1943.
14. Wasserman, D. P., *Advanced Methods in Neural Computing*. Van Nostrand Reinhold, Newyork, ISBN: 9780442004613, 255p., 1993.
15. Kennedy, J., and R. Eberhart, "Particle swarm optimization," *Proc. of IEEE International Conference on Neural Networks*, pp. 1942–1948, 1995.
16. Kern, D. J., and D. H. Werner, "A genetic algorithm approach to the design of ultra-thin electromagnetic bandgap absorbers," *Microwave and Optical Technology Letters*, vol. 38, no.1, pp. 61–64, Dec. 2003a.
17. Kern, D. J., D. H. Werner, M. J. Wilhelm and K.H. Church "Genetically engineered multiband high-impedance frequency selective surfaces," *Microwave and Optical Technology Letters*, vol. 38, no.5, pp. 400-403, Sep. 2003c.
18. Huang, R. F., Z. W. Li, L. B. Kong, L. Liu, and S. Matitsine, "Analysis and design of an ultra-thin metamaterial absorber," *Progress In Electromagnetics Research B*, Vol. 14, pp. 407–429, 2009.
19. Jiang, Z. H., Q. Wu, X. Wang, and D. H. Werner, "Flexible wide-angle polarization-insensitive mid-infrared metamaterial absorbers," *Proceedings of IEEE Antennas and Propagation Society International Symposium*, pp. 1–4, Jul. 2010.
20. Guodos, S. K., and J. N. Sahalos, "Microwave absorber optimal design using multi-objective particle swarm optimization," *Microwave and Optical technology letters*, vol. 48, no. 8, pp. 1553–1558, Aug. 2006.
21. Choudhury, B., S. Bisoy, and R. M. Jha, " Emerging trends in soft computing for metamaterial design and optimization," *Computers, Materials & Continua*, vol. 31, no.3, pp. 201–228, 2012.
22. Choudhury, B., S. Bisoy, and R. M. Jha, " Bacteria foraging algorithm for metamaterial design and optimization," *2013 IEEE Applied Electromagnetics conference (AEMC)*, Bhubaneswar, India, Paper No.: CAD-2-1884, 2P. Dec. 2013.
23. Holland, J. H., *Adaptation in natural and artificial system*, MIT Press, The university of Michigan press, 1975.

Chapter 2
Metamaterial-Based Miniaturized Planar Inverted-F Antenna

S. Manjula and Balamati Choudhury

2.1 Introduction

The fast-growing high-profile mobile communication systems demand miniaturization of mobile systems, which leads to the reduction in antenna size. The design of compact antennas with enhanced performance such as low specific absorption rate (c), large bandwidth and multiband operation is gaining momentum. Efficient use of metamaterial structure has got great importance in this miniaturization process.

Planar inverted-F antennas (PIFA) are widely used in mobile applications as it is compact, easy to fabricate, integrate into the printed circuitry and is light weighted. The inclusion of metamaterial structures to the PIFA system makes it more compact, which will overcome the challenges faced in the miniaturization of antenna systems in mobile communication. This chapter gives an insight to the possibilities of designing a compact metamaterial PIFA system.

2.1.1 PIFA System

The conventional PIFA is also known as quarter-wavelength patch antenna as it is resonating at quarter wavelength, i.e., at $\lambda/4$. Since the antenna is resonant at the quarter wavelength, the size of the antenna is less and hence occupying little space. PIFA systems have wide applications such as mobile hand sets. This chapter also provides a feasibility study of metamaterial PIFA system in the cellular bands. The frequency bands for cellular phone systems are as follows: GSM (Global System

S. Manjula · B. Choudhury (✉)
Centre for Electromagnetics, CSIR-National Aerospace Laboratories, Old Airport Road, Bengaluru 560017, India
e-mail: balamati@nal.res.in

B. Choudhury (ed.), *Metamaterial Inspired Electromagnetic Applications*,
DOI 10.1007/978-981-10-3836-5_2

for Mobile Communications: 880–960 MHz), DCS (digital communication system: 2.71–1.88 GHz), PCS (Personal Communication System: 1.85–1.99 GHz) and IMT-2000 (3G: 1.92–2.17 GHz).

Advantages of PIFA System

PIFA systems are chosen for its simple and compact structure. They can be easily installed in cellular devices. The main advantages of PIFA are mentioned below:

- The advances in printed circuit technology make antenna fabrication quite easier and cost-effective.
- As the size of the PIFA antenna is in the order of quarter wavelength, it found applications in UHF and VHF range.
- Enhanced antenna performance such as low SAR due to its reduced backward radiation.
- A single-patch PIFA system itself provides a maximum gain in the range of 6–9 dBi.
- The gain of the antenna can be enhanced by using an array of patches on a single substrate, which makes the antenna cost-effective as compared to other antennas.
- PIFA systems possess good electrical characteristics.
- It can be tuned easily, and impedance matching can be done by varying the position of the feed.
- It can be used for MIMO (multiple-input and multiple-output) applications.
- The losses of PIFA system are low as compared to other antennas.

Limitations

Planar inverted-F antenna is characterized by narrow bandwidth. The shorting pin known to reduce the size of the antenna will result in narrow impedance.

2.1.2 Applications of PIFA System

Planar inverted-F antenna finds applications in wireless applications such as cellular devices, biotelemetry communication, healthcare applications, WLAN and WIMAX applications for laptop. These applications require a much smaller antenna with a better performance and enhanced bandwidth.

Biotelemetry application: Recently, PIFA systems have been used in biotelemetry communication system for treatment of human diseases and also to monitor the signals within the human body. In biotelemetry systems, the antenna is implanted into the body. The body is continuously monitored by transmitting and receiving the physiological signals between implanted antenna and the medical communication device.

Radio frequency identification: Hospitals have advanced a lot in the recent years, by making use of the technologies available in the market to minimize medical errors and to provide better quality of healthcare in the hospitals. RFID tags which make use of PIFA system comes in handy by providing security solutions, and locating and tracking hospital staff in case of emergencies which is efficient for the hospitals to deliver a prompt service. RFID collars are also developed using PIFA system. The collars are usually worn by the cattle in a large herd. In case of theft or any other danger, the cattle will start running. The acceleration is detected by the device, and a message is sent to farmer through GSM network, making it easy to find the cattle.

WLAN/WIMAX applications: PIFA offers many attractive features such as small structure, low cost and better performance. These features have made it attractive in the field of wireless technology. PIFA serves as wireless local area network antenna for the laptop computers. In the recent trends, it has become a standard element in laptops for wireless internet access, due to its small structure. PIFA system is also used in cellular devices to provide wireless LAN services.

2.1.3 Design Challenges

Although there are various techniques and different simulation tools, the following issues make the design of PIFA system a truly challenging aspect.

Downsizing the antenna: The size of the antenna cannot be decreased just by scaling or by reducing the dimensions of the antenna since the impedance at antenna's terminals will be affected. The other factors which will be affected by reducing the size of the antenna are its radiation characteristics. Also, the gain, bandwidth and efficiency will be greatly influenced. The other limitation in reducing the size of the antenna is the difficulty in feeding small antennas correctly. Small antennas are greatly affected by their surroundings, so care should be taken to integrate such antennas into a system so as to obtain the maximum efficiency. Hence, we can say that reducing the size of antenna is to compromise the best bandwidth and efficiency among the obtained set of values. The other issues in design of antenna are discussed below.

Integrating with other components: While integrating an antenna onto the board of the mobile phones, the neighboring components has to be taken into consideration since they influence the performance of antenna greatly. With the advancement of multimedia components in cellular devices, the space for antenna started shrinking and antenna had to be integrated with other components. The cellular device consists of battery, display, speakers, camera and other components. These components act as noise source to the antenna since they emit RF noise which will degrade the performance of the antenna. Hence, care should be taken while integrating these components with the antenna so as to minimize its effect on the antenna performance and obtain maximum output performance.

Mechanical issues: Once the antenna is being designed electrically, the next step is to provide a mechanical solution to the electrical design. To maintain the efficiency of the antenna, it is necessary to have good mechanical contact and tolerance. Also, the distance between metal objects and lossy objects needs to be optimized for the better performance.

Quality issues: An antenna is quality-controlled through its design. To have a better quality, the antenna needs to be simple with few contact points. The distance between feeding pins and the ground pins needs to be large to improve tolerance.

Connection issues: Connection is another issue which needs to be considered to maintain the quality of the antenna. The antenna suffers ohmic losses due to the number of connections. This can be reduced by having direct connections to the PCB as much as possible.

2.2 Reviews on PIFA

A brief description about planar inverted-F antenna is given in this section. The concept of miniaturization, need for it and the challenges faced in miniaturization are discussed. Different techniques on miniaturization of the antenna are studied, and review is provided towards enhancing the gain of the antenna. Based on the survey, suitable technique for miniaturization is opted and is used in our design.

2.2.1 *Basic PIFA System*

Planar inverted-F antenna gets its name from its shape which is in the form of an inverted F. Figure 2.1 shows the schematic of a basic PIFA antenna consisting of a ground plane, substrate, feeding pin and a radiated patch shorted at one end through a shorting pin. The radiating patch is placed above the ground plane so as to be parallel with it. This in turn introduces capacitance to the input impedance of the PIFA system which is compensated by adding a short-circuit stub. When the PIFA

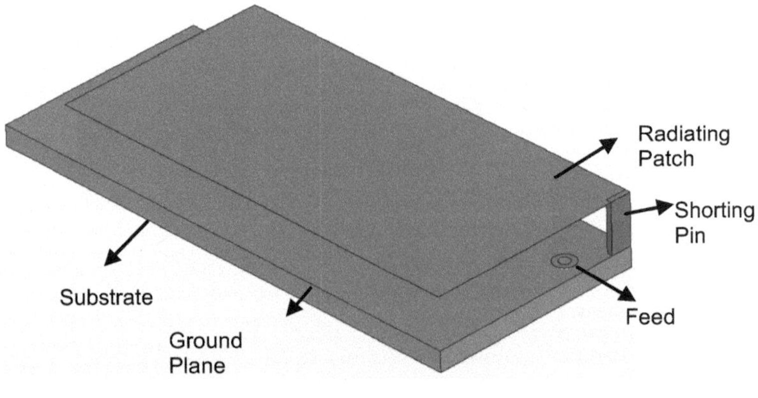

Fig. 2.1 Schematic of a conventional PIFA system

is excited, currents in the ground plane are also excited, which forms an electromagnetic field. The ground plane and the antenna combination acts as a perfect reflector only when the dimensions of the ground plane are larger than the antenna, i.e., the ground plane length required is $\lambda/4$, where λ is the operating wavelength. If the length of the ground plane is below $\lambda/4$, then the antenna possesses problems for the tuning mechanism, thereby reducing antenna's performance, while multi-lobes are introduced if the length of the ground plane is greater than $\lambda/4$. The feeding pin is placed in between the open end and the shorter end of the patch.

The input impedance of the PIFA can be controlled by varying the position of the feed. The impedance of the PIFA decreases when the feed is closer to the shorting pin, while the impedance increases when the feeding pin is moved away from the shorting pin. The impedance of PIFA can be tuned using this parameter. The resonant frequency of the PIFA system is dependent on the width "D" of the shorting wall. If the width of the patch is equal to the width of the shorting wall, then the shorting pin runs the entire width of the patch. The antenna has maximum radiation efficiency when

$$L = \lambda/4 \tag{2.1}$$

If the width of the short is zero, then the antenna resonates at

$$L + W = \lambda/4 \tag{2.2}$$

From the above two equations, it is seen that the antenna resonates when the space between the edge and the shorting area is a quarter wavelength. The resonant length of the antenna is approximated as

$$L + W - D = \lambda/4 \tag{2.3}$$

2.2.2 Miniaturization of PIFA

In these days, the size of the antenna in cellular phones is a major concern, as the space available for antenna in mobile phone is reducing. Hence, we primarily focused on reducing the size of the antenna while improving the gain of antenna with low SAR. Different techniques have been investigated for miniaturization of PIFA system and are given below.

Capacitive loading technique

Capacitive loading is the most common technique used to reduce the size of the conventional PIFA system. In this technique, capacitance is introduced between the open edge and feeding point as shown in Fig. 2.2. Such antennas are known as lumped element-loaded antennas.

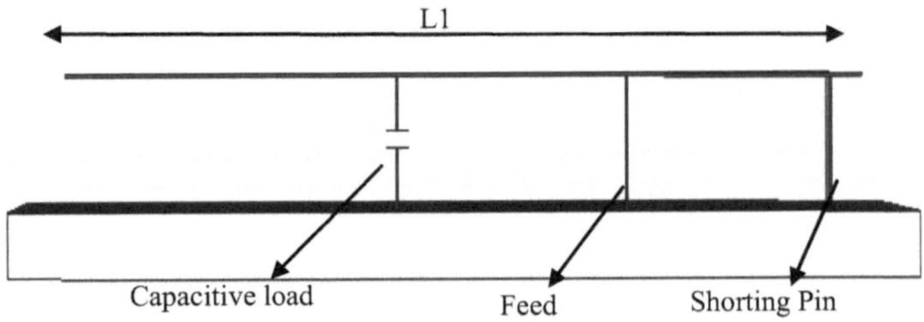

Fig. 2.2 Capacitive-loaded antenna

Chi et al. [1] proposed a miniaturized PIFA system by loading lumped elements achieved from transmission line model. The electrical size of the antenna is reduced to one-eleventh (reduction by 91%) after loading shunt capacitors periodically. The experimental results show that although the size is reduced significantly, the measured radiation gain and bandwidth are reduced to −22.6 dBi and 0.15%, respectively.

Magnetic material loading technique
Loading the antenna with magnetic materials is another known technique to reduce the size of the antenna. Kawano et al. [2] designed PIFA systems by loading magnetic materials in the feed and the shorting pin of the PIFA system. The corresponding input impedance, surface current distribution and the radiation characteristics of the antenna were obtained and compared with the calculated results. It is seen that the PIFA loaded with magnetic materials was 50% smaller in structure than the PIFA without magnetic materials.

Karkkainen et al. [3] used magnetic material fillings to reduce the size of the PIFA system. The PIFA system is designed using the dispersive magnetic material fillings between the ground plane and the radiating patch. The designed PIFA is simulated using FDTD method, and the effect of resonant frequency of the materials on the resonant frequency of the PIFA is studied. It has been observed that the resonance frequency of the PIFA is reduced by 20% due to the resonant magnetic material, while the relative bandwidth is retained.

Metamaterial-based PIFA system
In the recent years, metamaterials have inspired great deal of interest due to their unique properties such as negative permittivity and negative permeability. Metamaterials are not readily available in nature, and these materials are artificially engineered to obtain the negative characteristics by periodically arranging thin-wire structures. Split ring resonators on the other hand are also known to exhibit negative refractive index.

Xin-Yuan et al. [4] proposed a novel design of metamaterial-loaded wideband PIFA system. An array of Jerusalem cross-shaped PEC structure used in frequency-selective surface (FSS) is designed, which acts as a metamaterial structure. This metamaterial structure is loaded on the ground plane of the PIFA antenna. In metamaterial-loaded PIFA system, the size was reduced by 91%. The center frequency of the PIFA system is found to be 3.4 GHz.

EBG-based PIFA systems
The growing interest in electromagnetic band gap (EBG) structures paved way for researchers to utilize them in miniaturization of antenna. EBG structure helps in reducing surface waves, thereby enhancing the bandwidth, forward and backward radiation ratio, directivity and efficiency of the antenna.

Zuazola et al. [5] proposed a miniaturized PIFA system using FSS strips which act as EBG structures. FSS strips are incorporated on the bottom layer of the PIFA system to increase the bandwidth and front-to-back ratio. The frequency of operation of the FSS-based PIFA system was 1–6 GHz. PIFA with FSS strips mounted on a FR4 substrate having extra ground plane can broaden the bandwidth and is suitable for GSM 900 band. The miniaturized multiband PIFA with FSS strips can work satisfactorily at the DCS-1800, UMTS, Hiper LAN/2 bands and Bluetooth (WLAN).

Zhao et al. (2004) [6] proposed an EBG-based PIFA system to downsize the PIFA system. The designed antenna consisted of mushroom-like EBG structure operating from 2.1 to 2.4 GHz having metallic patches, connecting pins, a ground plane and a dielectric substrate. From the simulated results, it was seen that the PIFA with EBG had a reduced thickness of 2.6 mm as compared to the conventional PIFA.

2.2.3 Performance Enhancement of PIFA

Since years, there have been growing demands in the market for smaller antennas that can be easily embedded in cellular devices. This calls for the attention toward enhancing the performance of the antenna. Several researchers have been trying to find different methods to improve the performance of the antenna. Different techniques are discussed here to enhance the performance of the antenna.

Using electronic band gaps
EBG structures stand out as an option because of its unique properties and characteristics such as surface wave suppression and in-phase reflection coefficient. The surface wave suppression characteristic is often utilized to enhance the gain of the antenna and to reduce the mutual coupling, while the latter can be utilized to design low-profile antennas. EBG as metamaterials are used to control the propagation of waves at all incident angles and polarizations. EBG are also known as photonic crystals and are realized using periodic arrangement of metallic structures.

Zhao et al. (2004) [6] designed an EBG-based planar inverted-F antenna to increase the efficiency of the antenna. Mushroom-like EBG structure operating from 2.1 to 2.4 GHz having metallic patches, connecting pins, a ground plane and a dielectric substrate was used. EBG structures are capable of controlling the flow of electromagnetic waves. Compared to the conventional PIFA, PIFA with EBG structure has 6 dB reductions in backward radiation. The resonant frequency is shifted from 2.7 to 2.4 GHz.

EBG structure reduces surface waves, thus leading to an increase in directivity, bandwidth, forward and backward radiation ratios, and efficiency.

Abidin et al. [7] proposed a 2 × 2 U-shaped PIFA system at 2.4 GHz and reduced the mutual coupling effect using EBG structures. The designed antenna had a thin profile and was fabricated in FR4 substrate, which can be useful in compact mobile handsets. The EBG-based PIFA system reduced the mutual coupling by 6 dB.

Using metamaterials

Attia et al. [8] used metamaterials as superstrate to enhance the gain of a PIFA system working at UMTS band. Split ring resonators with high value of permeability at resonant frequency of the antenna were used as superstrate. A 12 × 12 array of metamaterial SRRs was used as superstrate on a conventional planar inverted-F antenna, which increased the antenna gain by 3.2 dB.

Xin-Yuan et al. [4] designed a metamaterial-loaded wideband PIFA antenna using frequency-selective surfaces (FSS). The FSS consisted of an array of Jerusalem cross-shaped PEC structure which acts as a metamaterial structure. This metamaterial structure is loaded on the ground plane of the PIFA system. The metamaterial-loaded PIFA system has an enhanced bandwidth of 49.4% from 2.56 to 4.24 GHz at a center frequency of 3.4 GHz.

To summarize, metamaterials play an important role in miniaturization and performance enhancement of PIFA system. The antenna size can be reduced significantly using metamaterial split ring resonators, metamaterial EBG structures and Jerusalem cross designs in the ground plane, to act as metamaterial. Hence, the literature survey reveals that metamaterial structure helps in miniaturization as well as performance enhancement of PIFA system.

2.3 Background Theories for the Design of PIFA System

This section provides an overview of the concepts used in design of PIFA system. As per our approach toward miniaturization, a basic PIFA system will be designed and the corresponding results for the designed antenna structure shall be obtained. The theory of metamaterials has been discussed, and a suitable metamaterial structure is selected for PIFA design. The selected metamaterial structure will be then optimized for desired resonant frequency. The designed metamaterial is loaded onto the basic antenna as per the proposed design, and the results are obtained. Further optimization of the novel PIFA has been carried out in order to achieve high system performance. The designed antenna with metamaterials is then compared with the results of the basic antenna structure, and its performance characteristics are studied.

2.3.1 Metamaterial Structure

From the literature survey, it could be seen that the metamaterial structure improves the performance of the PIFA system and also helps in downsizing the antenna. The most popular kind of metamaterial structure is the SRR. A single SRR can be modeled as an LC circuit. Figure 2.3 shows a simple square SRR. The SRR in the figure consists of a coil which forms an inductor, while the open ends of the coil form the capacitor.

The gaps between the rings are considered for the prediction of resonant frequency. Since the SRRs exhibit negative permeability close to their resonant frequency, it is important to predict the proper resonant frequency for a chosen structure.

Design of SRR

A split ring resonator is the most common structure used to obtain negative refractive index in designing metamaterials. SRRs were obtained initially from the wire structure by Sir Pendry [9]. SRRs can produce magnetic resonance leading to negative permeability. The return loss of antenna is tuned by magnetic resonance of the SRR which results in the improvement in size of the antenna. Further, the interaction between magnetic field of SRR and the fields generated by PIFA system will enhance the radiation properties, return loss, bandwidth and gain of antenna. The SRR structure considered here consists of two concentric square rings. Each ring has a gap appearing on opposite sides. Schematic of a simple square split ring resonator is shown in Fig. 2.3. In order to isolate the inclusions and their electric and magnetic properties, the non-planar concept was employed. The relative negative permittivity and permeability are given by

$$\varepsilon_r = n/\eta \tag{2.4}$$

$$\mu_r = n\eta \tag{2.5}$$

Fig. 2.3 A simple square split ring resonator

2.3.2 Specific Absorption Rate

The rate at which radiation is absorbed by the human body is measured by the specific absorption rate (SAR). It is usually measured over 1 gm/10 gm of human tissue. SAR is defined as

$$\mathrm{SAR} = \frac{\sigma E^2}{\rho}\,(\mathrm{W/kg}) \tag{2.6}$$

where E is the root-mean-square amplitude of induced field, σ is the conductivity of tissue, and ρ is the mass density of tissue. The radiation emitted from cellular devices is regarded as the top polluting sources by World Health Organization which may lead to health hazards in humans. Different countries around the globe have specified different standards to limit SAR produced from cellular devices. In USA, the specified SAR value is below 1.6 W/kg over 1 gm of tissue, while in Europe, the specified limit is around 2 W/kg over a tissue volume of 10 gm. India has adopted FCC standards to limit SAR. The value of SAR depends on volume of the head that has been exposed to radiation and also the position of cellular device to the head. For mobile applications, in order to reduce the electromagnetic interaction between human head and mobile phone, SAR has to be controlled. Various methods have been proposed towards reduction of SAR in human head.

Lin et al. [10] designed a FSS-based PIFA system towards reduction of SAR. The SAR was reduced by decreasing the backward radiation using two-dimensional periodic structure of frequency-selective surface (FSS) structure which was placed at the bottom side of the PIFA system. The resonating frequencies of the designed antenna are at 900 and 1800 MHz. The integration of PIFA and FSS also improves the antenna gain and radiation efficiency of the antenna. The total dimension of the PIFA system was reduced after addition of slotted FSS structure.

Manapati et al. [11] used metamaterial split resonators to reduce SAR of PIFA system. They proposed three different types of resonator such as spiral resonator (SR), split ring resonator (SRR) and open split ring resonator (OSRR) as metamaterial unit cell. The array of resonators exhibits negative permeability. The PIFA system which was resonating at 900 MHz was tilted by 30°, and the face of the antenna is in the direction opposite to the human head. By using the 10-element array of spiral resonator, split ring resonator and open split ring resonator, the SAR has been reduced by 35.52, 20.12 and 57.89%, respectively.

Faruque et al. [12] carried out electromagnetic absorption studies in mobile phones while using metamaterials in the antenna system. An array of metamaterials is kept in between the antenna and human head to reduce the SAR. A finite-difference time-domain (FDTD) method with lossy-Drude model was used for this analysis. In order to study the reduction in SAR, various parameters have been investigated, such as the effects of attaching the metamaterial at different positions, distance and size of metamaterials, perfect electric conductor and materials on the specific absorption rate reduction. The PIFA is constructed with helical-type antenna, which is resonating at 900 MHz, and is used for GSM applications. The metamaterial characteristics were obtained by arranging the split ring resonators periodically. The structure exhibits metamaterial properties at 900 and 1800 MHz. These metamaterial structures are designed on a printed circuit board for the easy integration with cell phones. Metamaterials helped to achieve a 42.12% reduction of the initial specific absorption rate value for 1 gm specific absorption rate and 53.94% reduction for the 10 gm specific absorption rate. Further, Faruque et al. [13] proposed a new design of square metamaterial structure and used as an array of metamaterials to reduce the absorption rate. The design was experimentally validated.

2.3.3 Optimization of PIFA Using PSO

The proposed PIFA system is designed using square SRR as metamaterials. The design of PIFA system involves varying the position of feed and changing the dimensions of shorting wall and radiating patch for better performance, and also the metamaterial needs to be designed to match the resonant frequency of the PIFA system which is a time-consuming task. Hence, optimization is done to obtain the required parameters for a good design. This calls for the implementation of different optimization techniques.

Soft computing technique stands as an option among various techniques due to their properties such as global optimization, quick response and accuracy. The emerging trends in soft computing techniques play an important role in the optimization of the design in various fields of electromagnetics including metamaterials. These techniques yield quick and robust solutions that are economically viable. Soft computing techniques have been earlier used to optimize the design of antennas and metamaterials.

Here, a square SRR as metamaterial in conjunction with PIFA system is used to downsize the antenna in order to enhance the performance of it. To accomplish this, an array of square SRRs is loaded on the antenna by replacing the ground plane. The SRR array needs to be designed at a frequency range which is equivalent to the resonant frequency of the PIFA system. To design the SRR at a particular resonant frequency, particle swarm optimization (PSO) is developed.

2.4 Design Methodology

PIFA systems are widely used in wireless communications due to their compactness and multiple-input and multiple-output applications. However, these antennas posses a narrow bandwidth which is a disadvantage and also the size of the antenna is to be taken into consideration. In order to increase the performance of the antenna and also to reduce its size, a simple square ring resonator is proposed along with conventional PIFA. The square ring resonator is to be designed at a resonant frequency equivalent to the resonant frequency of the PIFA system. PSO algorithm is used to optimize the structural parameters of the SRR for the desired resonant frequency. The optimized design of SRR is used in miniaturization of the antenna and to enhance the performance. The steps below are followed to design PIFA system.

Step 1: Design of a conventional planar inverted-F antenna using transmission line model (for the required resonant frequency).

Step 2: Selection of metamaterial structures and development of equivalent circuit model.

Step 3: Optimization of the selected metamaterial structures for the desired resonant frequency using particle swarm optimization (CAD package developed by CEM, CSIR-NAL).

Step 4: Design and simulation of optimized metamaterial structures. Using these metamaterial configurations, the PIFA system will be designed. The implementation of metamaterial structures as substrate or superstrate or hybrid ground plane depends on the EM requirements and application.

Step 5: To achieve high performance, optimization of the metamaterial PIFA system will be carried out using soft computing techniques.

Step 6: A comparative study of performance parameters of conventional PIFA system with metamaterial PIFA system will be carried out.

2.4.1 *Design of a Conventional PIFA System in C-Band*

The aim of this work is to downsize the PIFA system using metamaterials. As the first step, we design the basic PIFA system by interfacing FEM-based electromagnetic solvers with MATLAB and study its characteristics so as to compare its results with the metamaterial-loaded antenna. The structure of the conventional PIFA is shown in Fig. 2.4.

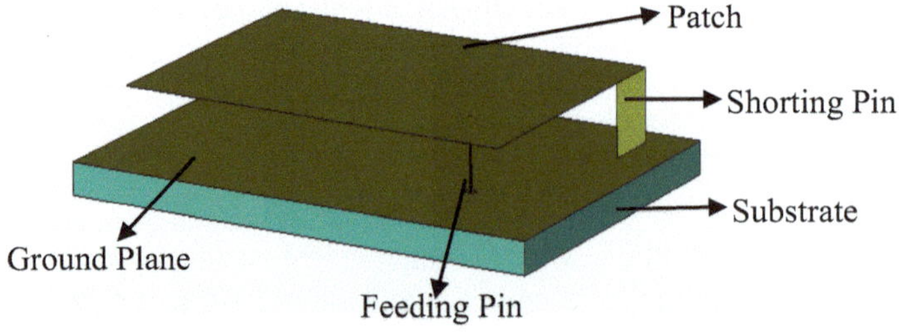

Fig. 2.4 Structure of PIFA without metamaterials

It consists of a ground plane, radiating patch, substrate, shorting pin and a feeding pin. The substrate has a dielectric constant of 2.55. Copper material is used for the radiating patch, shorting pin, feed and the ground plane. The proposed metamaterial-loaded antenna design is shown in Fig. 2.5. The return loss characteristics are shown in Fig. 2.6. Also Figs. 2.7, 2.8 and 2.9 demonstrate the radiation characteristics obtained for the conventional PIFA system. Figure 2.10 shows the graphical user interface for the design of metamaterial SRR and Table 2.1 provides the design parameters for the PIFA system.

The frequency sweep is set from 1 to 8 GHz. The center frequency is taken at minimum return loss, and the bandwidth can be calculated from return loss

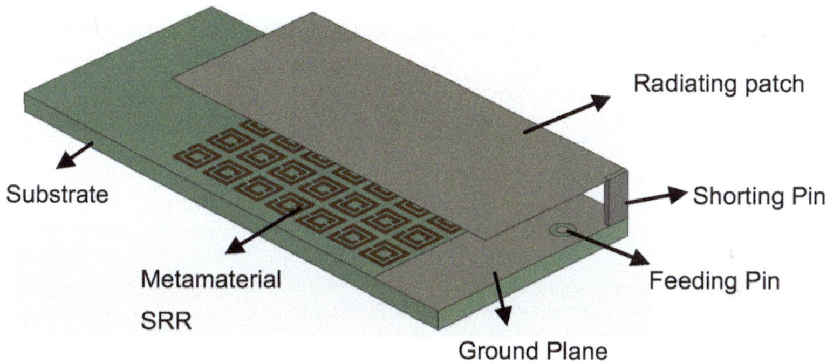

Fig. 2.5 Structure of proposed PIFA design with metamaterial

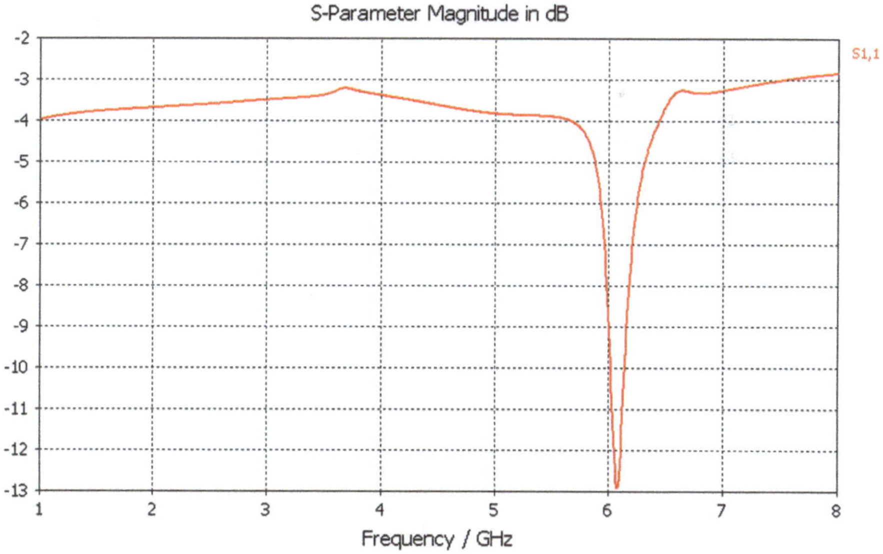

Fig. 2.6 Return loss characteristics of conventional PIFA structure

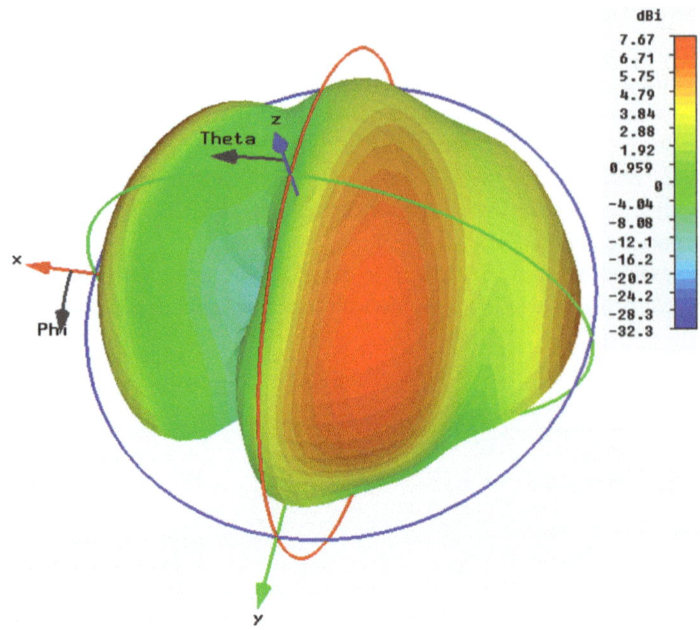

Fig. 2.7 Radiation pattern of the conventional PIFA structure

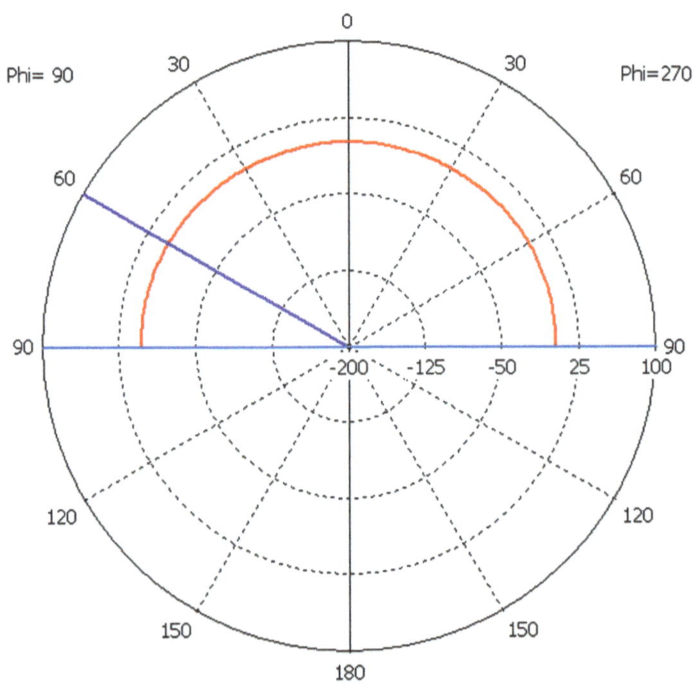

Fig. 2.8 Impedance plot (phi = 90) of conventional PIFA structure

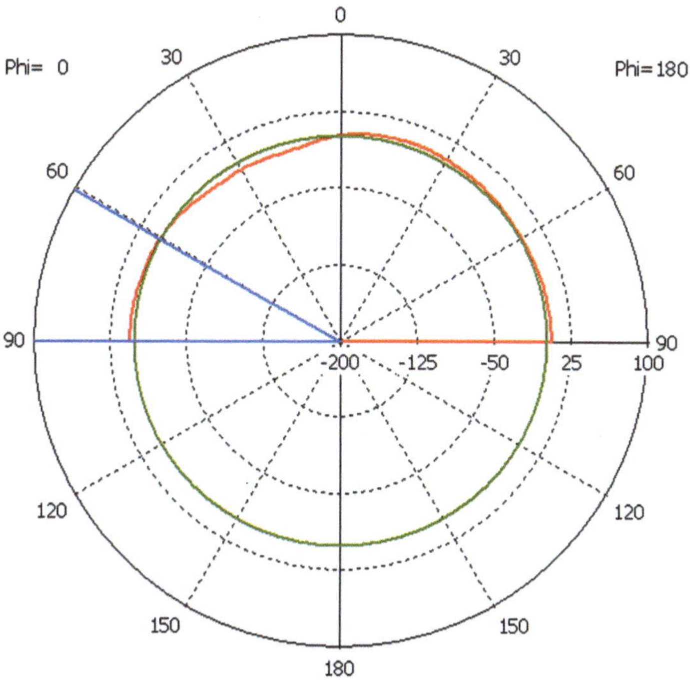

Fig. 2.9 Polar plot (phi = 0) of the conventional PIFA structure

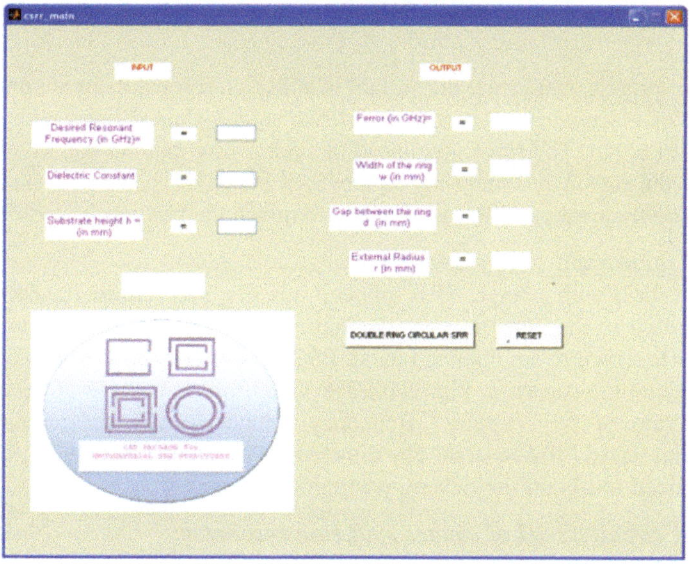

Fig. 2.10 Graphical user interface of PSO-based CAD package for design of metamaterial structures

Table 2.1 Dimensions of conventional PIFA system

Parameter	Value (mm)
Width of the substrate, W	25
Length of the substrate, L	40
Distance of the feed in x-axis, f_x	5.5
Distance of feed in y-axis, f_y	5.6
Width of the radiating patch, W_r	20
Length of the radiating patch, L_r	32
Height of the patch from ground, h	3.5
Thickness of the substrate, T	2

Table 2.2 Performance parameters of conventional PIFA system

Frequency	6.075 GHz
S_{11}	−12.893 dB
Gain	3.474 dB
Directivity	7.670 dBi

characteristics. The bandwidth of the antenna is taken for the frequencies above which the return loss is more than 10 dB. For the designed dimensions, the antenna is resonating at 6.075 GHz with return loss of 12.893 dB. The bandwidth of the antenna is found to be 23.28 MHz with center frequency of 6.075 GHz. The obtained antenna performances are shown in Table 2.2.

2.4.2 *Design of Metamaterials*

Square split ring resonators are the variant of split ring resonators used commonly in the design of metamaterials. It consists of two square-shaped rings with each ring consisting of a gap. The rings are placed in such a way that the gap on one ring is opposite to the gap on the other ring. Square SRR at 6 GHz is chosen as metamaterial unit cell. The frequency of SRR is designed to match the frequency of PIFA system.

Design of square split ring resonator

The design starts with a single SRR unit cell initially, and further, the SRR array is used according to the antenna size. For the preliminary design, structural dimensions of SRR structure are obtained using PSO-based CAD tool. The GUI window of the CAD tool is shown in Fig. 2.10. The CAD tool is capable of designing the SRR structure at any resonant frequency which is a priori requirement in metamaterial-based PIFA system. The dimensions calculated from the CAD tool are then optimized to obtain the desired resonant frequency.

Equivalent circuit model of square split ring resonator

The square SRR can be modeled as LC parallel tank circuit. The gap in the ring forms the capacitor, while the metal ring forms the inductor. The values of capacitance and inductance of the split ring resonators are determined using the

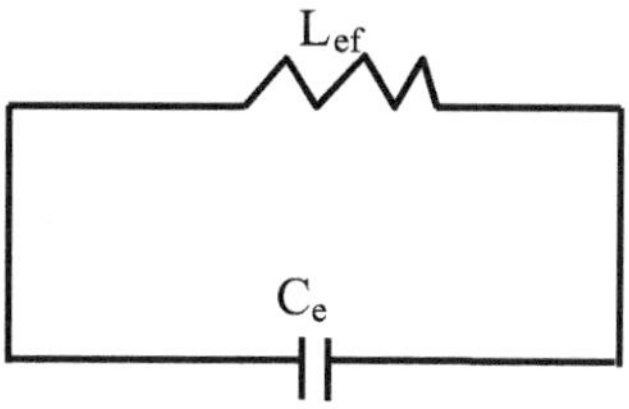

Fig. 2.11 Equivalent circuit of square split ring resonator

equivalent circuit shown in Fig. 2.11. From the equivalent circuit analysis, resonant frequency of the SRR is given by

$$w_0 = \frac{1}{2\pi\sqrt{L_{ts}C_{eqs}}} \tag{2.7}$$

where C_{eqs} is the total equivalent capacitance and L_{ts} is the total effective inductance of the SRR.

The total equivalent capacitance is given by

$$C_{eqs} = \left(a - \frac{3}{2}(w+d)\right) C_{pul} \tag{2.8}$$

where C_{pul} is the per-unit-length capacitance between the rings.

$$C_{pul} = \varepsilon_0\, \varepsilon_{eff} \frac{K\left(\sqrt{1-k^2}\right)}{K(k)} \tag{2.9}$$

and ε_{eff} is the effective dielectric constant given by

$$\varepsilon_{eff} = \frac{\varepsilon_r + 1}{2} \tag{2.10}$$

$K(k)$ denotes the complete elliptical integral of the first kind with k denoted as

$$k = \frac{d}{d+2w} \tag{2.11}$$

The total effective inductance of the square SRR is given by

$$L_{ts} = \frac{4.86\,\mu_0}{2}(a-w-d)\left[\ln\left(\frac{0.98}{\rho}\right) + 1.84\rho\right] \tag{2.12}$$

where ρ is the filling factor given by

$$\rho = \frac{w+d}{a-w-d} \tag{2.13}$$

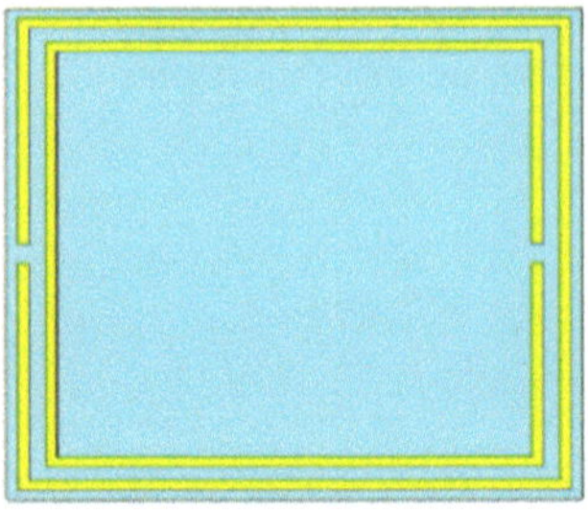

Fig. 2.12 Optimized design of square SRR

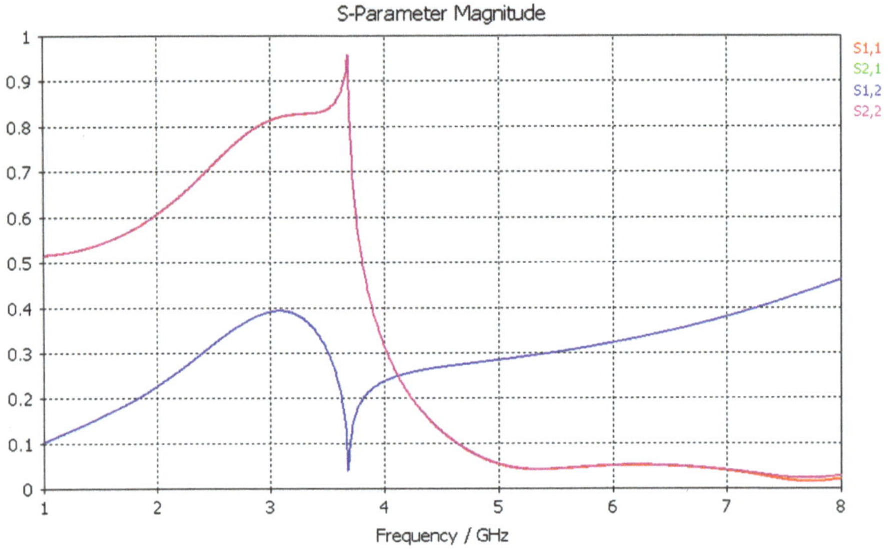

Fig. 2.13 Scattering parameters of square SRR

Simulation of square split ring resonator

Numerical simulations are used to predict the transmission properties of SRRs. Here FEM-based solver is used to simulate the SRR structure. The simulations are done by propagating EM wave along Y-direction. Z-axis was electrically polarized, and X-axis was magnetically polarized. The background material was assumed to be vacuum. To validate the simulation, the SRR parameters were chosen to be $d = 0.1204$ mm, $g = 0.2408$ mm, $c = 0.6995$ mm and $l = 5.2$ mm. The thickness and dielectric constant of the substrate were 2 and 2.55 mm, respectively. Table 2.3 provides the structural parameters of the metamaterial unit cell. The structure was then optimized for resonant frequency of 6 GHz, and the corresponding permittivity and permeability were extracted. Figure 2.12 shows the optimized structure of square SRR, and the simulation results are shown in Figs. 2.13, 2.14 and 2.15. From the simulation results, it has been observed that the designed square SRR metamaterial has negative material characteristics from 4 to 6 GHz.

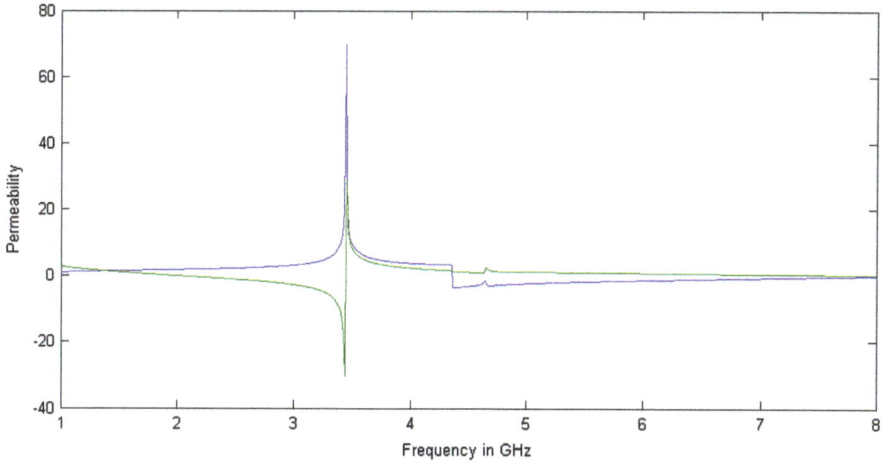

Fig. 2.14 Extracted permittivity of square SRR

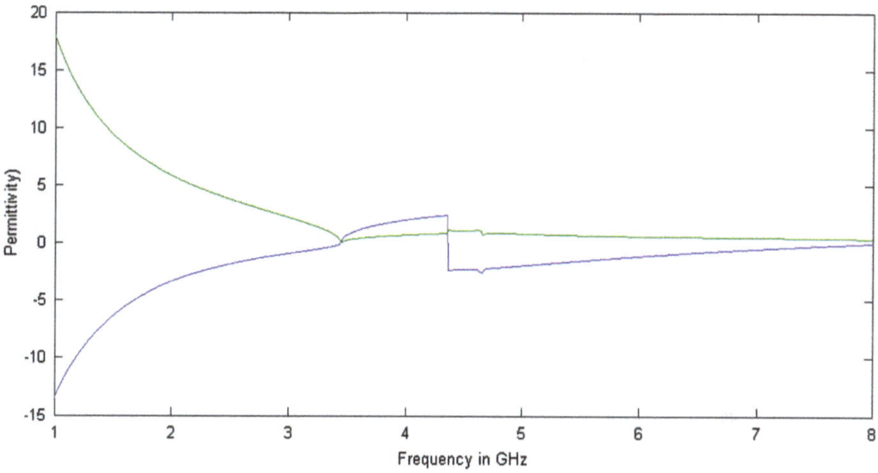

Fig. 2.15 Extracted permeability of square SRR

Table 2.3 Dimensions of designed metamaterial

Parameter	Value (mm)
Thickness of the substrate, *h*	2
Metal thickness, *t*	0.01
Width of the rings, *w*	0.6995
Length of outer ring, *o*	5.2
Gap in each ring, *g*	0.2408
Gap between rings, *d*	0.1204

2.4.3 Design of PIFA Antenna Using SRR

The design of PIFA with metamaterials is shown in Fig. 2.16. An array of SRRs is taken and placed on the substrate by replacing the ground plane. The antenna is then scaled accordingly to resonate at 6 GHz frequency. The design optimization and simulations are done by interfacing FEM-based EM solver with MATLAB. The structural dimensions of PIFA system using SRR has been optimized for better performance using soft computing based optimization algorithm in conjunction with EM solvers. The enhanced antenna characteristics are shown in Figs. 2.17, 2.18, 2.19, 2.20 and 2.21.

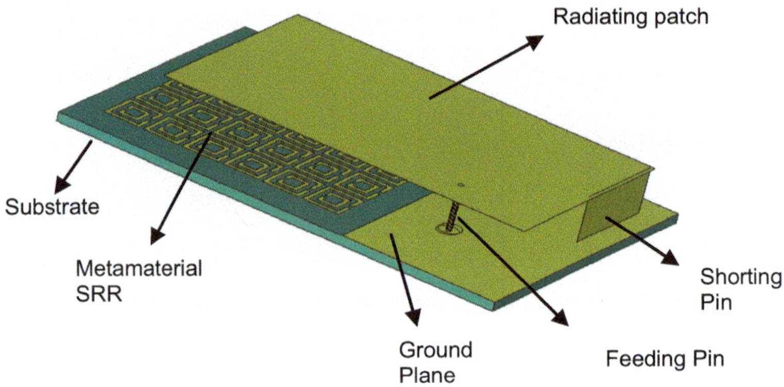

Fig. 2.16 Structure of PIFA design with metamaterials

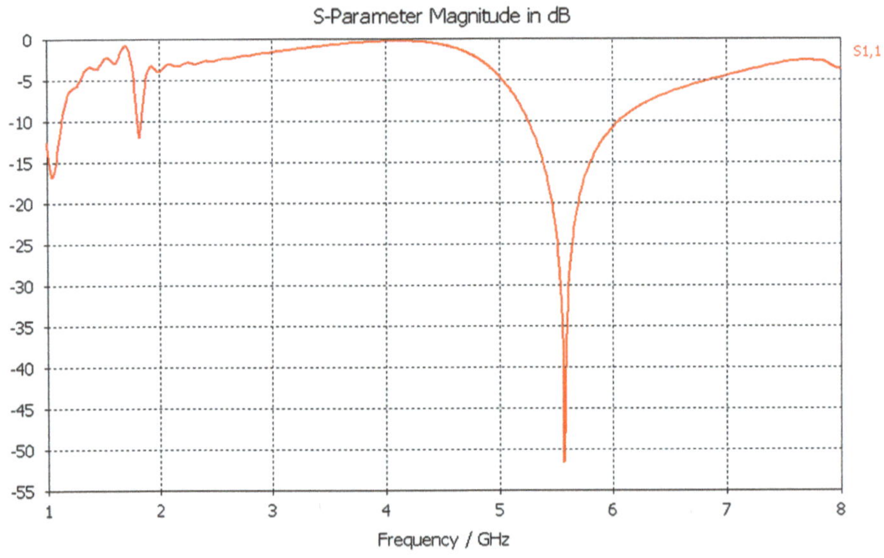

Fig. 2.17 Return loss characteristics of PIFA system with metamaterials

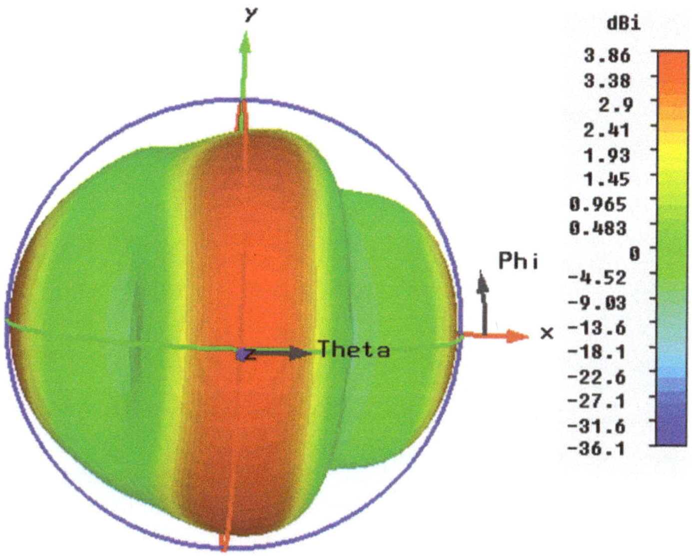

Fig. 2.18 3-D radiation pattern (directivity) of PIFA system

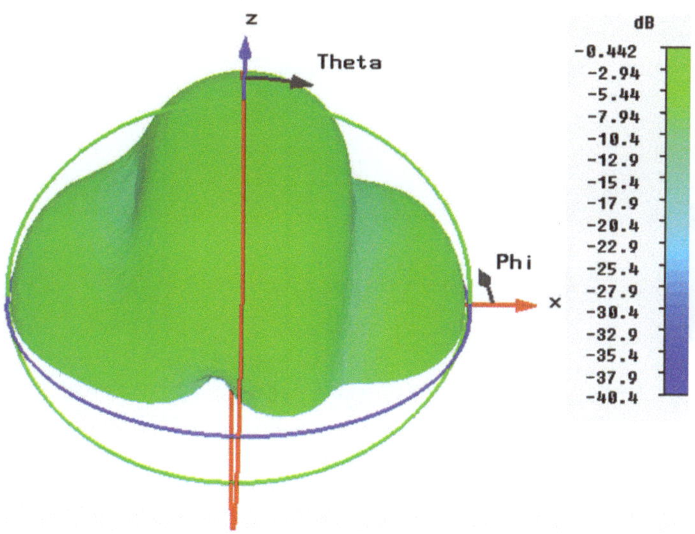

Fig. 2.19 3-D radiation pattern (gain) of PIFA system

The designed PIFA system with metamaterials resonates at a frequency of 5.564 GHz with return loss of 51.57 dB. The directivity at resonant frequency is 3.86 dBi. Table 2.4 provides the obtained antenna characteristics. From the results, it is clear that the return loss, gain and directivity of the metamaterial-loaded PIFA

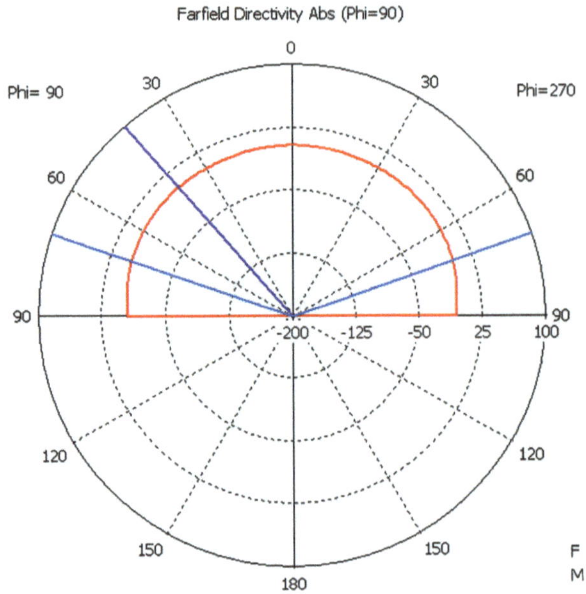

Fig. 2.20 Polar plot (phi = 90) of PIFA system

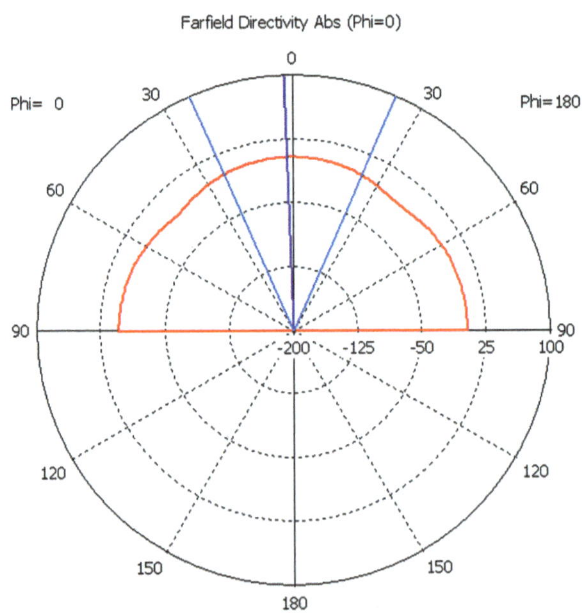

Fig. 2.21 Polar plot (phi = 0) of PIFA system

Table 2.4 Performance parameters of metamaterial-based PIFA system

Frequency	5.564 GHz
S_{11}	−51.57 dB
Directivity	3.86 dBi

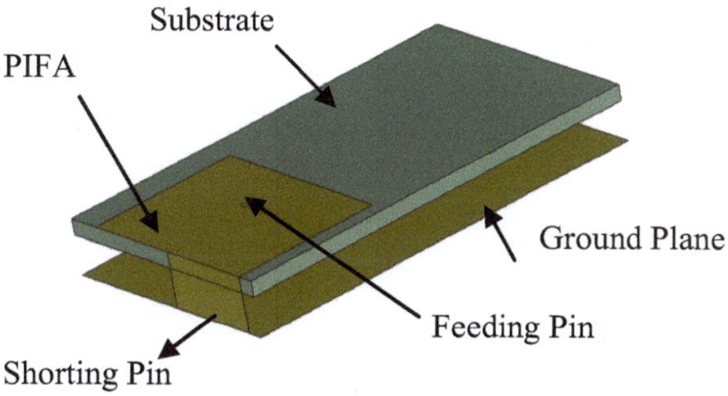

Fig. 2.22 Schematic of a conventional PIFA system

antenna have been improved as compared to a conventional PIFA system without metamaterial inclusion.

2.4.4 *Design of PIFA Antenna Using PBG Substrate in S-Band*

The efficient antenna system can be achieved by enhancing bandwidth and downsizing the system. The PIFA with PBG structure will give admirable results compared to the conventional PIFA. In recent times, the PBG structure has been developed rapidly. The operation of the PBG mainly depends on periodicity, pattern, dielectric contrast and repetitive spacing between atoms. These structures do not let the propagation of electromagnetic waves for certain space directions in a given frequency range, which is very constructive feature. In this section, a novel PBG planar inverted-F antenna (PIFA) for wearable systems is presented.

The structure of the conventional PIFA is shown in Fig. 2.22.

It consists of a patch, ground plane, shorting wall, feeding point and Teflon substrate. The thickness of the substrate is 2 mm with permittivity $\varepsilon_r = 2.1$. The PIFA patch is manufactured on Teflon substrate. The shorting wall having length 25 mm and width 24 mm is placed to achieve the impedance matching. There is an air gap between substrate and ground plane, it will not change the field of antenna, but it can change the resonant frequency. The height of air gap is 6 mm. Copper material is used for radiating patch, shorting pin, feed and ground plane. The FEM-based solver is used to simulate the PIFA system. The detailed dimensions of conventional PIFA are given in Table 2.5, and the simulation results are shown in Figs. 2.23, 2.24, 2.25, 2.26.

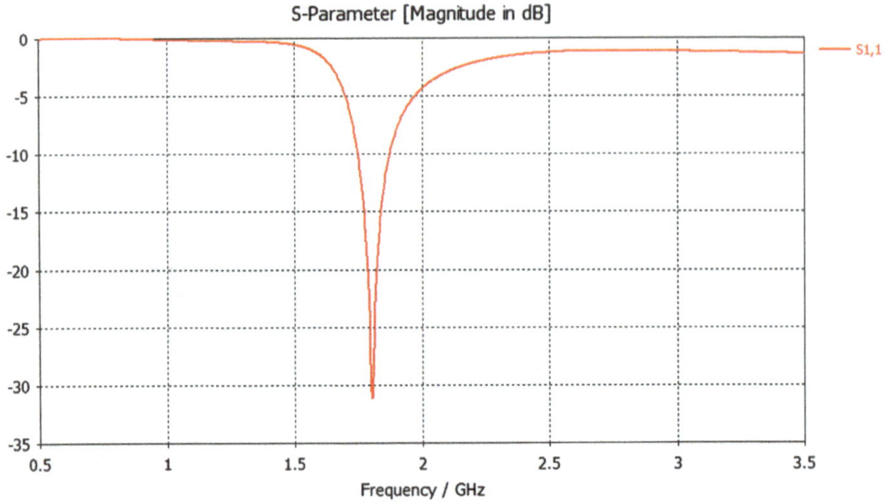

Fig. 2.23 Return loss (S11) for the conventional PIFA structure

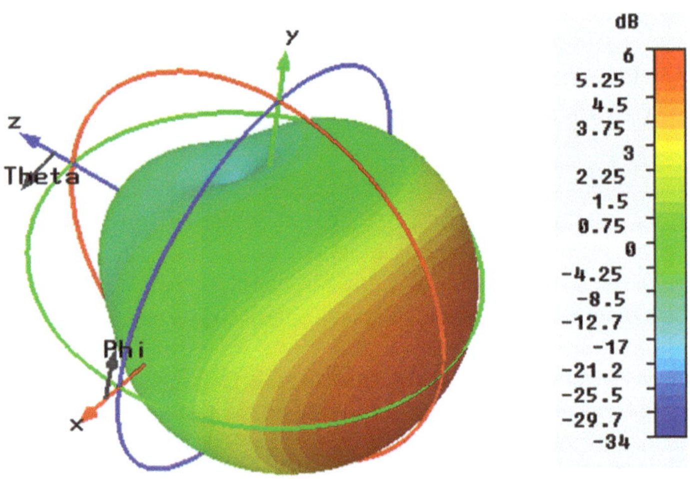

Fig. 2.24 Radiation pattern of the conventional PIFA structure

The frequency sweep is set from 0.5 to 3 GHz. For the designed dimensions, the antenna is resonating at 1.805 GHz with a return loss of 13.708851 dB. The standard procedure is followed to obtain the impedance bandwidth (VSWR < 2) of 25.4%. Table 2.6 gives the performance parameters of the conventional PIFA systems.

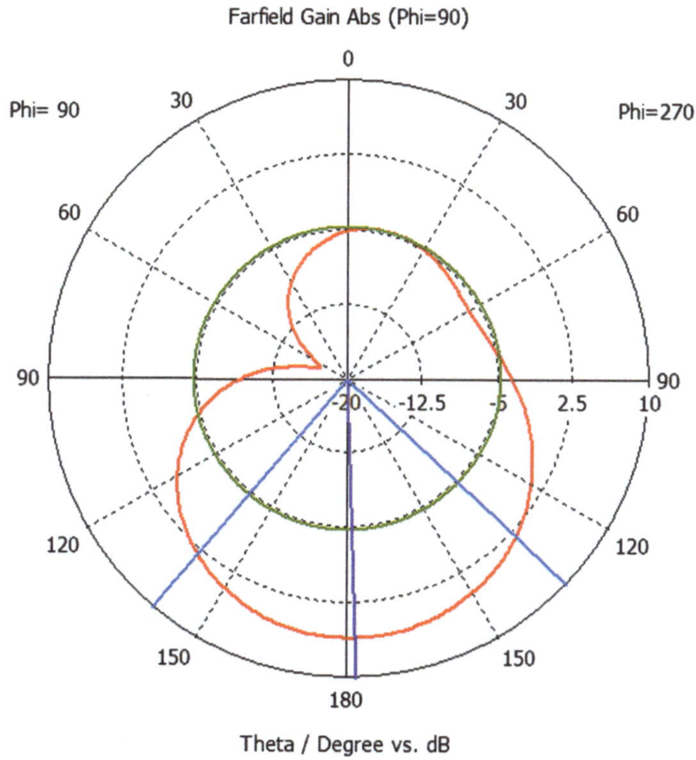

Fig. 2.25 3-D polar plot (phi = 90) of the conventional PIFA structure

The design of PIFA with PBG structure is shown in Fig. 2.27. The PBG structure in this section is formed by some cylindrical air gaps, which are fixed on the Teflon substrate. The PBG structure has been designed using cylindrical air gaps over a dielectric substrate in a hexagonal pattern. Although the cylindrical air gaps has been placed periodically, the feed probe area has been kept blank i.e. without cylindrical air gaps to avoid the effect of feed probe. The dimension of the designed cylindrical air gap has diameter d = 3 mm and periodicity i.e. the distance between cylindrical patch is a = 5 mm. The finite element method (FEM)-based EM solver is used for simulation of PBG-based PIFA. The design and simulation results are shown in Figs. 2.28, 2.29, 2.30 and 2.31.

The PBG-based antenna's return loss characteristics are shown in Fig. 2.28, the antenna is resonating at 1.853 GHz. The standard procedure has been followed to obtain the impedance bandwidth (VSWR < 2) of 39.9%. The obtained performance parameters for the PIFA system with PBG structure are given in Table 2.7. The simulated impedance bandwidth of the novel PBG-based PIFA is improved 14% than that of the conventional PIFA. The gain of the novel PBG PIFA is increased slightly.

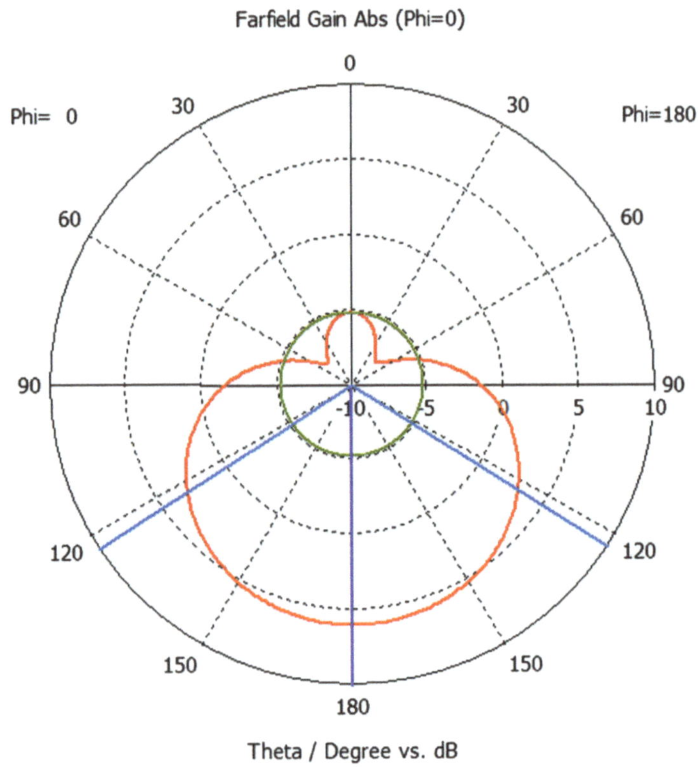

Fig. 2.26 3-D polar plot (phi = 0) of the conventional PIFA structure

Fig. 2.27 Schematic of a PIFA system with PBG substrate

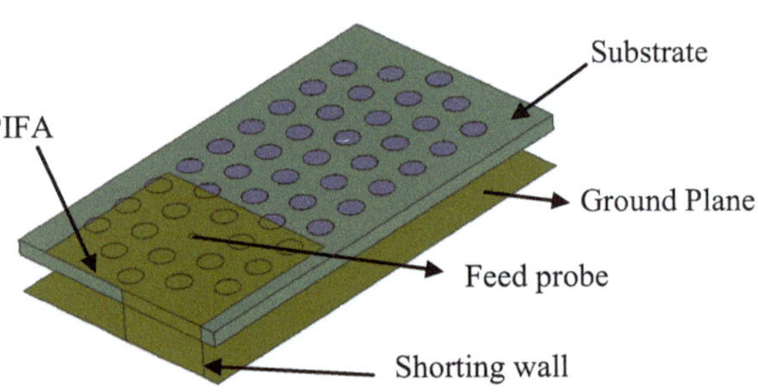

2.4.5 Procedure to Calculate Impedance Bandwidth

The performance of the antenna will be affected by high efficiency and large bandwidth parameters, which are interrelated to each other, and one does not have full freedom to separately set these parameters. The performance of an electrically small antenna can be examined in the same way as a simple resonator with the

Table 2.5 Dimensions of PBG-based PIFA system

Parameter	Value (mm)
Width of the substrate	30
Length of the substrate, L	50
Width of the patch, w	24
Length of the patch, l	25
Distance of the feed in x-axis, f_x	0
Distance of feed in y-axis, f_y	15
Height of the substrate from ground, h	6
Thickness of the substrate, T	2

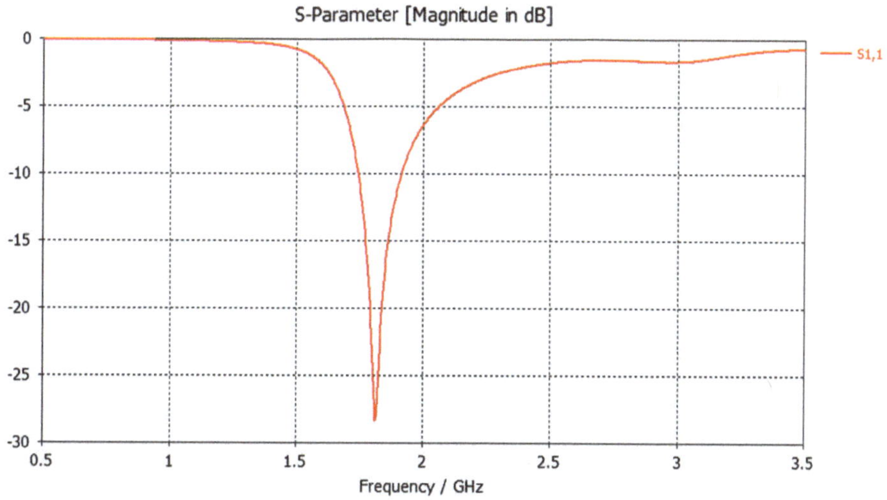

Fig. 2.28 Return loss for the PIFA structure with PBG

Table 2.6 Performance parameters of conventional PIFA system

Frequency	1.805 GHz
S_{11}	−13.70885 dB
Gain	5.871 dB
Directivity	5.851 dBi
Impedance bandwidth	25.4%

resonance frequency f_r and the quality factor Q. There are two losses in antenna, represented by external quality factor Q_e and unloaded quality factor Q_u. The total losses of antenna is represented by Q_l and is given by

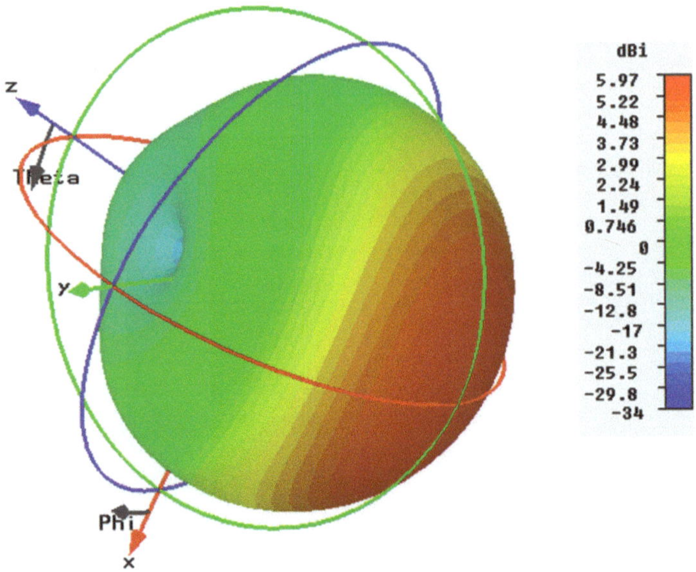

Fig. 2.29 Radiation pattern of PIFA structure with PBG

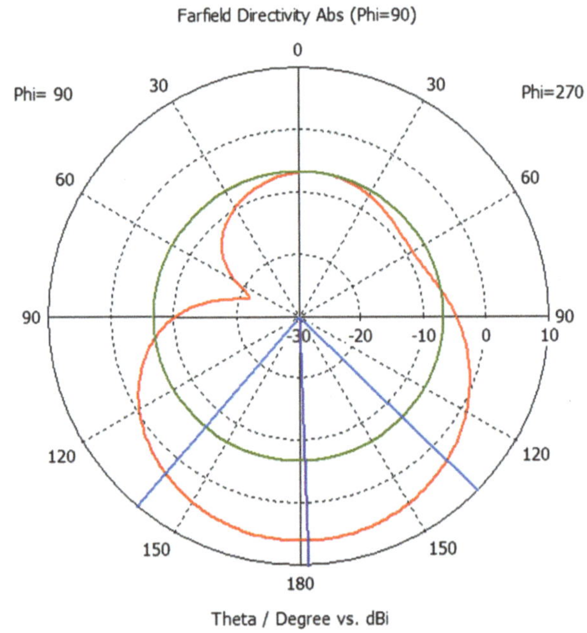

Fig. 2.30 Radiation pattern (in polar plot (phi = 90)) of the PIFA structure with PBG

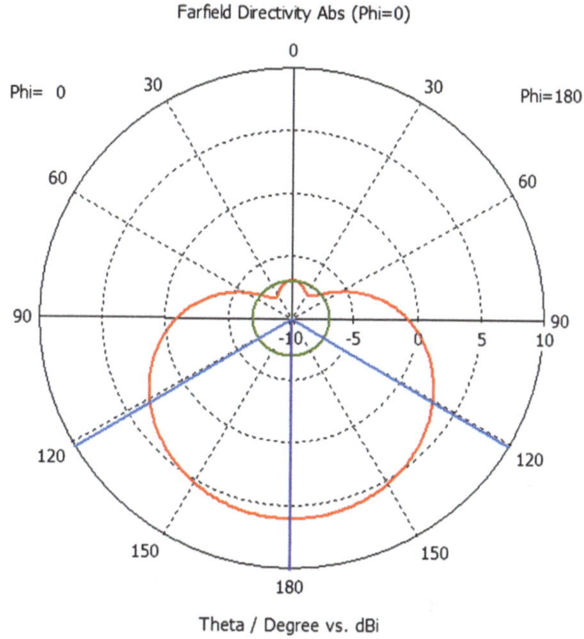

Fig. 2.31 Polar plot (phi = 0) of the PIFA structure with PBG

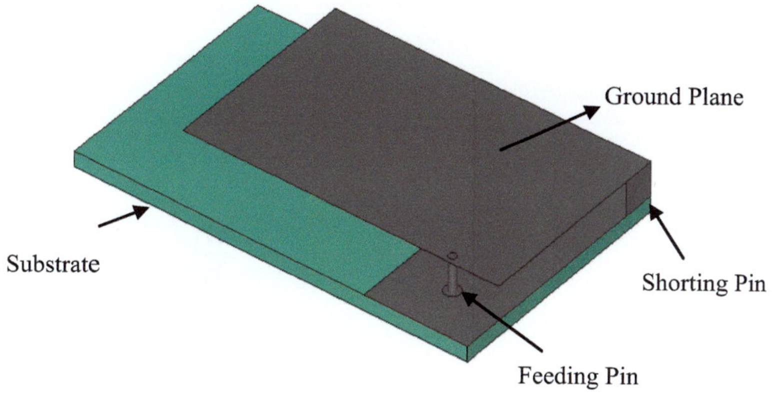

Fig. 2.32 Schematic of conventional PIFA

$$Q_l = \frac{f_r}{B_{hp}} \tag{2.14}$$

where B_{hp} is the half-power bandwidth. The unloaded quality factor can be calculated by

Table 2.7 Performance parameters of PBG-based PIFA system

Frequency	1.853 GHz
Return loss	−12.71 dB
Gain	5.990 dB
Directivity	5.971 dBi
Impedance bandwidth	39.9%

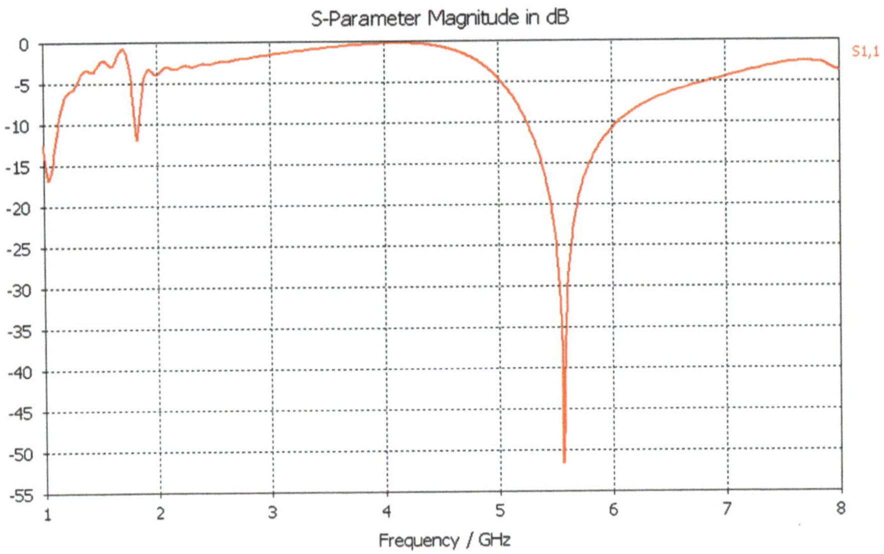

Fig. 2.33 Return loss characteristics of a conventional PIFA

Fig. 2.34 Radiation pattern of the conventional PIFA structure

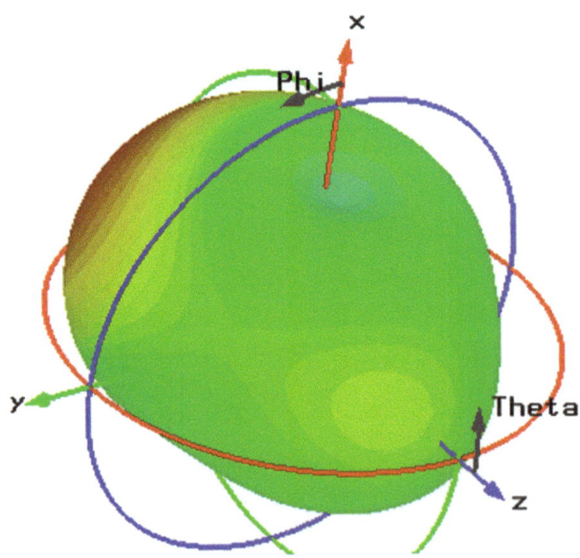

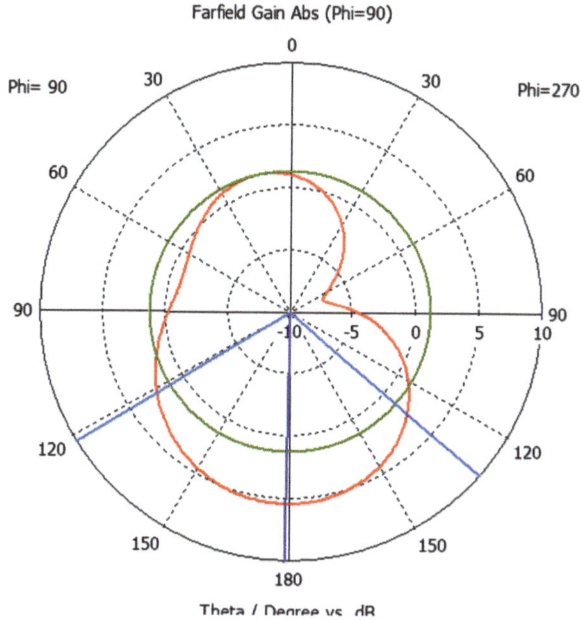

Fig. 2.35 Radiation pattern (polar plot (phi = 90)) of the conventional PIFA structure with metamaterials

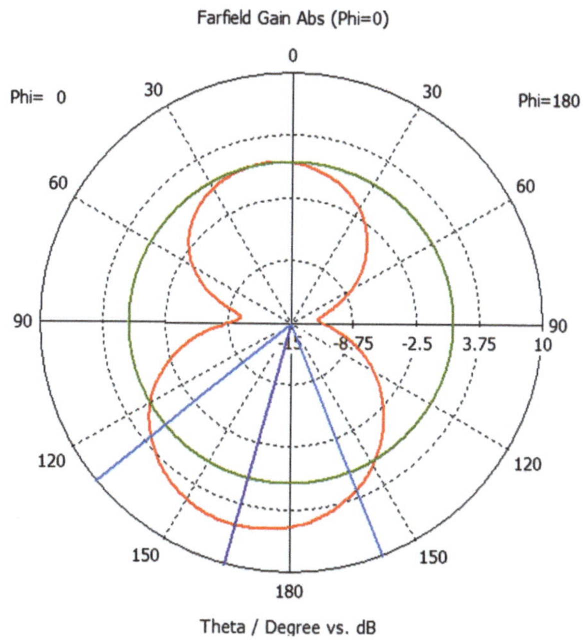

Fig. 2.36 Radiation pattern (polar plot (phi = 0)) of the conventional PIFA structure with metamaterials

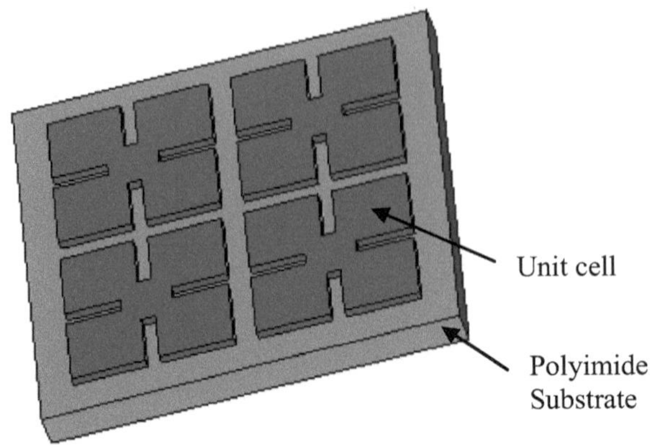

Fig. 2.37 Design of metamaterial unit cell

Table 2.8 Dimensions of conventional PIFA system

Parameter	Value (mm)
Width of the substrate	30
Length of the substrate, L	50
Width of the patch, w	40
Length of the patch, l	25
Distance of the feed in x-axis, f_x	6
Distance of feed in y-axis, f_y	6.5
Height of the substrate from ground, h	4.5
Thickness of the substrate, T	2

$$Q_u = \frac{2 * Q_l}{1 \pm \rho(f_r)} \tag{2.15}$$

where the plus or minus sign refers to under- or over-coupled antenna, respectively, and $\rho(f_r)$ is the reflection coefficient at resonance frequency. At critical coupling, Eq. (2.15) is reduced to

$$Q_u = 2 * Q_l \tag{2.16}$$

The unloaded quality factor is a physical constant and specific to each antenna structure. Thereby, it is not affected by coupling, and the impedance bandwidth for matching criterion VSWR $\leq S$ can be given by

$$\text{Bandwidth(BW)} = \frac{S - 1}{Q_u * \sqrt{S}} \tag{2.17}$$

This equation assumes critical coupling although the maximum bandwidth is obtained with a slight over-coupling. In Eq. (2.17), S is the standing wave ratio and assumed as 2. The measured voltage standing wave ratio should be less than S.

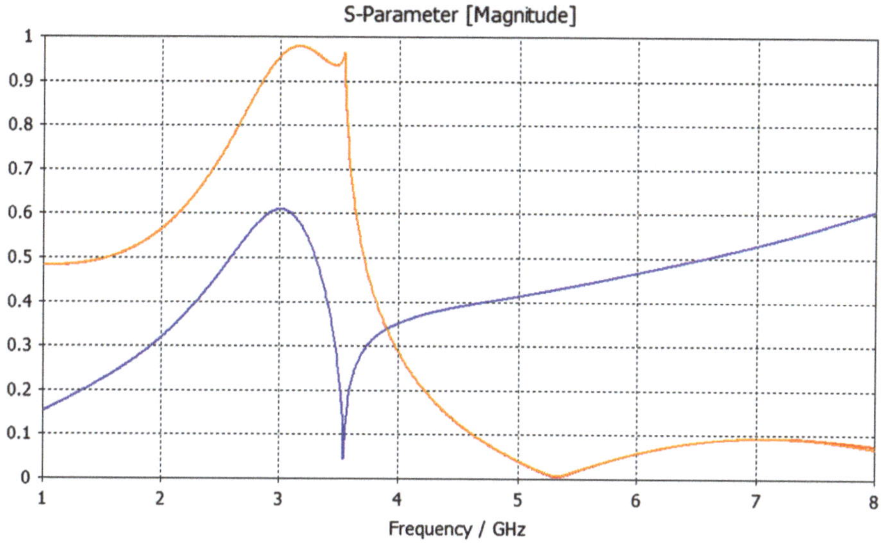

Fig. 2.38 Scattering parameters of SRR

Table 2.9 Performance parameters of conventional PIFA system

Frequency	5.564 GHz
S_{11}	−51.57 dB
Gain	3.252 dB
Directivity	4.670 dBi

The main intention behind the PIFA system with metamaterials is to downsize the antenna system. The FEM-based solver is used to design the antenna system and study the characteristics of a metamaterial-loaded antenna so as to compare its results with the conventional PIFA. The structure of the regular PIFA is shown in Fig. 2.32.

The planar inverted-F antenna (PIFA) consists of a substrate, radiating patch, ground plane shorting wall and feed. The dielectric constant of a substrate is 2.25. Copper material is used as the radiating patch, shorting wall, feed and ground plane. The dimensions of the substrate are 50 × 30 mm, respectively. The thickness of the polyimide substrate is 2 mm. The length and width of the ground plane are 13 and 30 mm, respectively. The height of the shorting wall is 4.5 mm and is placed in order to achieve impedance matching. The radiating patch is placed above the ground plane having dimensions of 40 and 25 mm, respectively. The FEM-based simulation software is used to simulate the PIFA. The feed has to be placed in a proper position in such a way that it obtains good results. The detailed dimensions of the conventional PIFA are given in Table 2.8, and the simulation results are shown in Figs. 2.33, 2.34, 2.35 and 2.36.

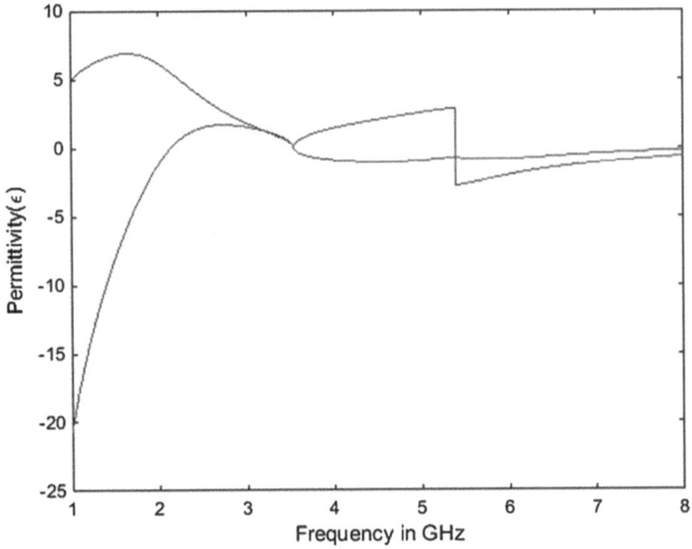

Fig. 2.39 Permittivity characteristics of SRR

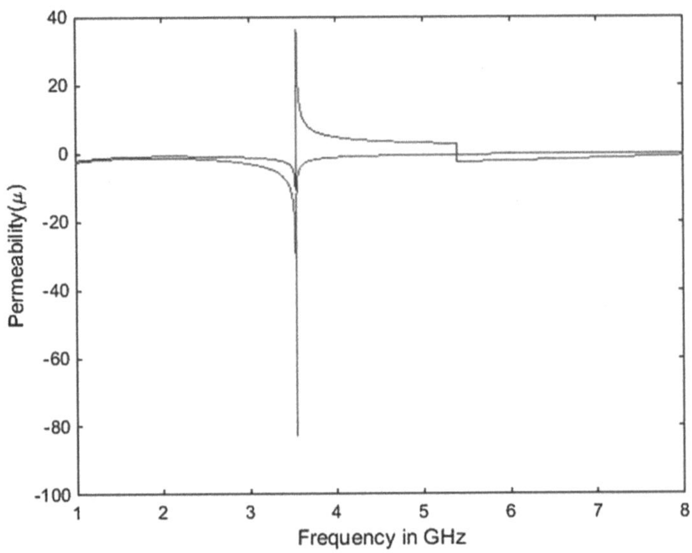

Fig. 2.40 Permeability characteristics of SRR

The operating frequency range of PIFA system is 1–8 GHz. The center frequency is taken at minimum return loss, and the bandwidth can be calculated from return loss. The reflection coefficient of −51.57 dB is obtained at a resonant

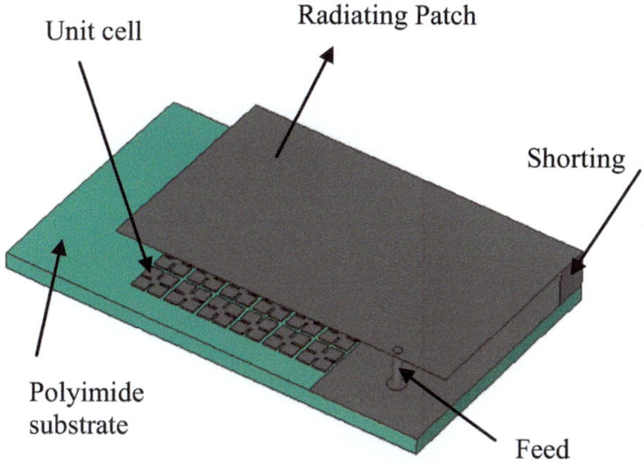

Fig. 2.41 Schematic of PIFA with metamaterial unit cell

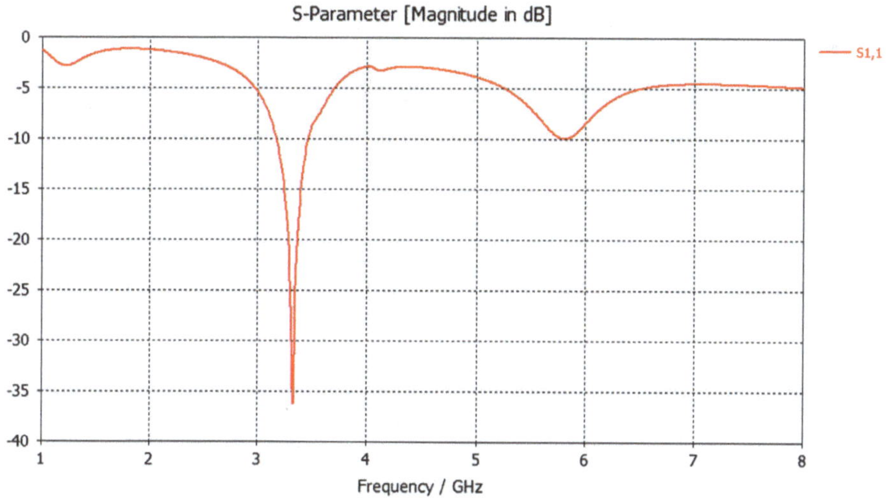

Fig. 2.42 Return loss characteristics of PIFA with metamaterial unit cell

frequency of 5.564 GHz for the designed dimensions of the antenna. The bandwidth of the antenna is found to be 23.28 MHz. The corresponding gain and directivity are 3.252 and 4.670 dBi. The obtained performance parameters for the conventional PIFA system are given in Table 2.9.

Numerical simulations are used to predict the transmission properties of SRRs. Here FEM-based solver is used to simulate the SRR structure. The simulations are done by propagating EM wave along z-direction. Vacuum was chosen as the background material. To validate the simulation, the SRR parameters were chosen

Table 2.10 Dimensions of metamaterial unit cell

Parameter	Value (mm)
Height of the substrate, *h*	2
Metal thickness, *t*	0.2
Width of the unit cell	7.5
Length of unit cell	7.5
Gap in each part of unit cell, *g*	0.3
Gap between unit cell, *d*	0.5

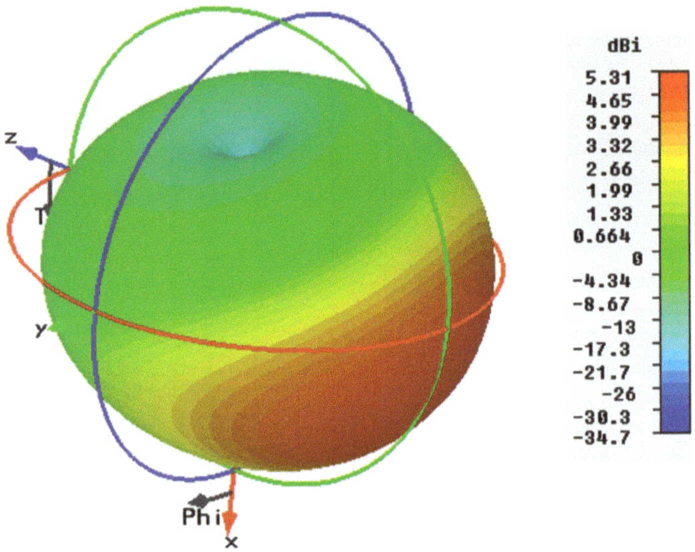

Fig. 2.43 Radiation pattern of PIFA with metamaterial unit cell

to be $d = 1.2$ mm, $g = 0.3$ mm, $w = 7.5$ mm and $l = 7.5$ mm. The thickness and dielectric constant of the substrate were 0.75 and 2.23 mm, respectively. The thickness of unit cell is 0.2 mm. The unit cell is made up of copper material. The structure was then optimized for resonant frequency of 3 GHz, and the corresponding permittivity and permeability were extracted. Figure 2.37 shows the optimized structure of SRR, and the simulation results are shown in Figs. 2.38, 2.39 and 2.40. From the simulation results, it was seen that the designed square SRR metamaterial has negative parameters from 4 to 6 GHz. The dimensions of the designed metamaterial unit cell is given in Table 2.10.

The corresponding permittivity and permeability characteristics calculated using return loss are given below.

The design of PIFA with metamaterial unit cell is shown in Fig. 2.41. An array of SRRs is taken and placed on the substrate. The antenna is then scaled

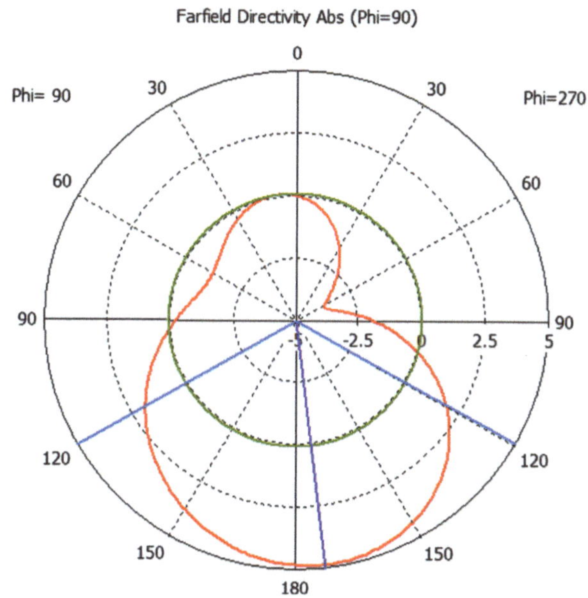

Fig. 2.44 Radiation pattern (polar plot (phi = 90)) of PIFA structure with metamaterials

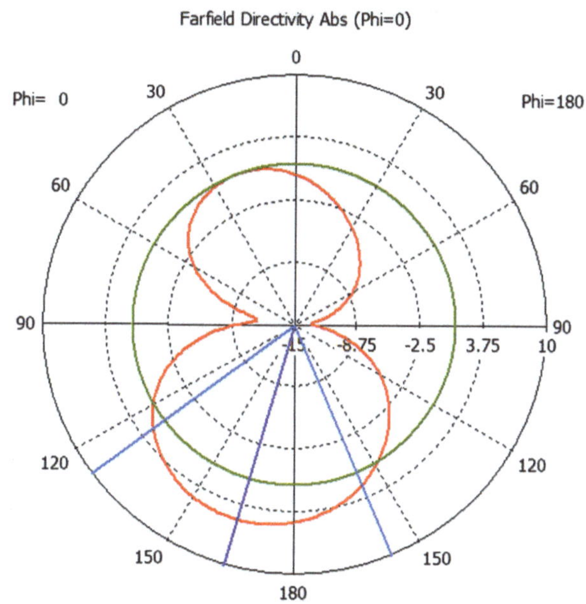

Fig. 2.45 Radiation pattern (polar plot (phi = 0)) of the PIFA structure with metamaterials

accordingly to resonate between 3 and 4 GHz frequencies. The design and simulations are done by FEM-based solver. The design and corresponding results are shown in Figs. 2.42, 2.43, 2.44 and 2.45.

Table 2.11 Performance parameters of PIFA system with metamaterial unit cell

Frequency	3.324 GHz
S_{11}	−36.21248 dB
Gain	5.323 dB
Directivity	5.315 dBi

From simulation results, it can be shown that the PIFA system with metamaterials resonates at 3.324 GHz having reflection coefficient of −36.21 dB. The gain and directivity at 5.564 GHz are 5.323 dB and 5.315 dBi, respectively.

The deigned PIFA system without metamaterials resonates at a frequency of 5.564 GHz with return loss of 51.57 dB. The return loss, gain and directivity of PIFA with metamaterials are improved as compared to the conventional PIFA.

2.5 Conclusion

Planar inverted-F antennas are widely used in mobile systems because of its compact size and easy fabrication. These PIFA systems also find its application as wireless sensors that can be used in aircraft fuel tanks to check the fuel availability status. A technical review of advancements in design optimization of PIFA systems has been reported in this chapter. It has been observed that the antenna size can be further reduced by introducing metamaterial structures. In view of this, a metamaterial-loaded compact planar inverted-F antenna has been designed and simulated. The designed scattering parameter S_{11} of the metamaterial-based PIFA antenna is −51.57 dB, whereas that of the conventional PIFA is −12.893 dB only.

References

1. Chi, P. L., R. Waterhouse, and T. Itoh, "Antenna miniaturization using slow wave enhancement factor from loaded transmission line models," *IEEE Transactions on Antennas and Propagation*, vol. 59, no. 1, pp. 48–59, Jan. 2011.
2. Kawano,Y., S. Hayashida, S. Bae, Y. Koyanagi and H. Morishita, "A study on miniaturization of 900 MHz and 2 GHz band antennas utilizing magnetic material," *2005 IEEE Antennas and Propagation Society International Symposium*, pp. 347–350 vol. 3B, 2005.
3. Karkkainen, M., S. Tretyakov, and P. Ikonen, "Numerical study of a PIFA with dispersive material fillings," *Microwave and Optical Technology Letters*, vol. 45, no. 1, pp. 5–8, Apr. 2005.
4. Xin-Yuan, L., F.J. Huil, Z. Kuang, H. Jun, and W. Qunl, "A compact wideband planar inverted-F antenna (PIFA) loaded with metamaterial," *Proceedings of IEEE Cross Strait Quad-Regional Radio Science and Wireless Technology Conference*, pp. 549–551, Jul. 2011.
5. Zuazola, I.J.G., J.C. Bachelor, and R.J. Langley, "Miniaturised multiband PIFA antenna using a frequency selective surface (FSS) ground plane," *Proceedings of IEEE Lough borough Antennas and Propagation Conference*, pp. 281–284, Apr. 2003.

6. Zhao, Y., Y. Hao, and C. G. Parini, "Radiation properties of PIFA on electromagnetic bandgap substrates," *Microwave and Optical Technology Letters*, vol. 44, no. 1, pp. 21–24, Jan. 2011.
7. Abidin Z. Z., Y. Ma, R. A. Abd-Alhameed, K. N. Ramli, D. Zhou, M. S. Bin-Melha, J. M. Noras, and R. Halliwell, "Design of 2 x 2 U-shape MIMO slot antennas with EBG material for mobile handset applications," *Progress in Electromagnetic Research, PIERS Online*, vol. 7, no. 1, pp. 81–84, 2011.
8. Attia, H., M. M. B. Suwailam, and O. M. Ramahi, "Enhanced gain planar inverted-F antenna with metamaterial superstrate for UMTS applications," *Proceedings of Progress in Electromagnetics Research Symposium*, pp. 494–497, Jul. 2010.
9. Pendry, J.B.. A.J. Holden, W.J. Stewart, and I. Youngs, Extreme Low Frequency Plasmons in Metallic Nano structures.
10. Lin, H. N., K. W. Lin, and S. C. Chen, "Use of frequency selective surface to prevent SAR and improve antenna performance of cellular phones," *Proceedings of Electromagnetics Research Symposium*, pp. 214–218, Sep. 2011.
11. Manapati, M. B. and R. S. Kshetrimayum, "SAR reduction in human head from mobile phone radiation using single negative metamaterials," *Journal of Electromagnetic Waves and Applications*, vol. 23, no. 10, pp. 1385–1395, 2009.
12. Faruque, M. R. I., M. T. Islam, and M. A. M. Ali, "A new design of metamaterials for SAR reduction," *Measurement Science Review*, vol. 13, no. 2, pp. 214–218, 2013.
13. Faruque, M. R. I., M. T. Islam, and N. Misran, "Analysis of electromagnetic absorption in mobile phones using metamaterials," *Electromagnetics*, vol. 31, no. 3, pp. 215–232, 2011.

Suggested Bibiliography

1. Baek S. and S. Lim, "Low-profile automotive antenna for omnidirectional vertical polarized signal reception," *Electromagnetics,* vol. 32, no. 8, pp. 481–494, 2012.
2. Chen, H. T., K. L. Wong, and T. W. Chiou, "PIFA with a meandered and folded patch for the dual-band mobile phone application," *IEEE Transactions on Antennas and Propagation*, vol. 51, no. 9, pp. 2468–2471, Sep. 2003.
3. Chen, J. H., Y. L. Ban, H. M. Yuan, and Y. J. Wu, "Printed coupled-fed PIFA for seven-band GSM/UMTS/LTE WWAN mobile phone," *Journal of Electromagnetic Waves and Applications,* vol. 26, pp. 390–401, 2012.
4. Dwivedi V., Y. P Kosta, R. Jyoti, and V. T. Patel, "An investigation on design and application issues of miniaturized compact microstrip patch antennas for RF wireless communication systems using metamaterials: A study," *Proceedings of IEEE International RF and Microwave Conference,* pp. 226–231, 2008.
5. Hao, Y., Y. Zhao, Y. J. Lee, and I. J Youngs, "Electrically small antennas with dielectric, magnetodielectric and metamaterial loading," *Proceedings of IEEE Loughborough Antennas and Propagation Conference,* pp. 57–62, Apr. 2007.
6. Ikonen, P. M. T., K. N. Rozanov, A. V. Osipov, P. Alitalo, and S. A. Tretyakov, "Magnetodielectric substrates in antenna miniaturization: Potential and limitations," *IEEE Transactions on Antennas and Propagation*, vol. 54, no. 11, pp. 3391–3399, Nov. 2006.
7. Jalal, A. S. A., A. Ismail, A. R. H. Alhawari, M. F. A. Rasid, N. K. Noordin, and M. A. Mahdi, "Miniaturized metal mount Minkowski fractal RFID TAG antenna with complementary split ring resonator," *Progress in Electromagnetics Research C,* vol. 39, pp. 25–36, 2013.
8. Jasik.H., (ed.), Antenna Engineering Handbook, New York, McGraw Hill, 1961.
9. Karkkainen, M., S. Tretyakov, and P. Ikonen, "Numerical study of a PIFA with dispersive material fillings," *Microwave and Optical Technology Letters,* vol. 45, no. 1, pp. 5–8, Apr. 2005.

10. Kulkarni, S. D. and S. N. Makarov, "A compact dual-band foam-based UHF PIFA," *Proceedings of IEEE Antennas and Propagation Symposium*, pp. 3609–3612, doi:10.1109/APS.2006, Jul. 2006.
11. Lee, C. J., K. M. K. H. Leong, and T. Itoh, "Compact dual-band antenna using an anisotropic metamaterial," *Proceedings of The 36th European Microwave Conference*, pp. 1044–1047, 2006.
12. Lo.Y.T. and S.W. Lee, Antenna Handbook New York, Van Nostrand Reinhold, 1998.
13. Narayan S., Shamala J. B., R.U. Nair, and R.M. Jha, "Electromagnetic performance analysis of novel multi-band metamaterial FSS for millimeter wave applications," *TechScience Press Journal CMC: Computers, Materials & Continua*, vol. 729, p. 1–8, 2012. (Invited Paper).
14. Negative Refraction Metamaterials, Fundamental Principles and Applications. Wiley Interscience, A John Wiley and Sons, Inc., Publication.
15. Pintos, J. F., A. Louzir, P. Minard, J. Perraudeau, J. L. Mattei, D. Souriou, P. Queffelec, "Ultra-miniature UHF antenna using magneto-dielectric material," *IEEE International Symposium on Antenna Technology and Applied Electromagnetics and the American Electromagnetics Conference,* 4p., 2010.
16. Saraereh, O. A., M. Jayawardene, P. McEvoy, and J. C. Vardaxoglou, "Simulation and experimental SAR and efficiency study for a dual-band PIFA handset antenna (GSM 900/DCS 1800) at varied distances from a phantom head," *Technical Seminar on Antenna Measurements and SAR (AMS)*, pp. 5–8, doi:10.1049/ic:20040066, 2004.
17. Staub. O., J-F. Zurcher, A. Skrivervik, "Some considerations on the Correct Measurement of the Gain and Bandwidth of Electrically Small Antennas, "Microwave and Optical Technology Letter, 17,3,February 201998,pp. 156–160.
18. Xiong, J., Y.F. Yu, Y.M. Liu, and X. Geng, "An electrically small planar loop antenna with high efficiency for mobile terminal applications," *Journal of Electromagnetic Waves and Applications*, vol. 26, pp. 744–756, 2012.

Chapter 3
Electromagnetic Perspective of Tensors

Susan Thomas and Balamati Choudhury

3.1 Introduction

The computational electromagnetics for planar applications generally deals with scalars and vectors, whereas the conformal analysis which is an essential requirement in aerospace platform relies on the electromagnetic analysis beyond scalars and vectors, i.e., the tensors of rank 2. The emerging field of metamaterial science and technology into aerospace domain [1] is forcing researchers to carry out electromagnetic analysis of higher ranks. Hence, the fundamental concepts of scalars, vectors and tensors from an electromagnetic perspective are very much essential for the design and analysis of aerospace applications such as conformal antenna analysis, radar cross section (RCS) reduction by cloaking structures [2] and aerodynamic stress.

The circuit theory deals with the terminal voltage across a conductor and current through a conductor, whereas electromagnetic theory gives the concept of the electric and magnetic fields interacting in the space between the conductors. This chapter focuses on the fundamental terminologies such as scalar, vector and tensors from electromagnetic (EM) perspective. The EM parameters that are generally used in computational electromagnetics in aerospace domain are discussed elaborately by categorizing to the scalars, vectors and tensors of higher-order ranks.

S. Thomas (✉)
School of ECE, REVA University, Bangalore 560064, India
e-mail: susanthomas@reva.edu.in

B. Choudhury
Centre for Electromagnetics, CSIR-National Aerospace Laboratories,
Bangalore 560017, India

B. Choudhury (ed.), *Metamaterial Inspired Electromagnetic Applications*,
DOI 10.1007/978-981-10-3836-5_3

3.2 Tensors: A Mathematical Perspective

Tensors in general can be classified based on their rank or order or degree. The fundamental mathematical concepts on tensors have been discussed in this section.

3.2.1 *Tensors of Rank 0*

Scalar is a physical quantity that is represented using a dimensional number that indicates the magnitude of that quantity at a point in space and time. It is a tensor of rank zero. The general matrix representation of a scalar can be given by a 1×1 matrix, i.e., a matrix having only one component.

3.2.2 *Tensors of Rank 1*

Any physical quantity having both magnitude and direction is called as a vector. It is a tensor of rank 1. The general matrix representation can be given by $1 \times n$ row matrix or $n \times 1$ column matrix where 'n' denotes the dimensionality of vector. A two-dimensional vector ($n = 2$) has two components, which can be described as

$$v = v_1x_1 + v_2x_2 = \sum v_ix_i \tag{3.1}$$

Similarly, a three-dimensional vector ($n = 3$) has three components which can be expressed as

$$v = v_1x_1 + v_2x_2 + v_3x_3 = \sum v_ix_i \tag{3.2}$$

Here x_1, x_2 and x_3 indicate the components of a vector 'v' or are called as basis vectors. The vector 'v' has three components and three basis vectors. The vector can represent the direction of any physical quantity. Also a vector can represent a surface area by a normal to that surface.

3.2.3 *Tensors of Rank Greater Than 1*

Tensors are mathematical objects that can be used to represent real-world systems. The physical quantities whose analysis requires a track of the amplitude, direction and the plane on which component acts are known as tensors. A tensor contains the information about the directions and the magnitudes in those directions. Scalar is a zeroth-order tensor. Vector is a first-order tensor. Tensors of order/ranks greater

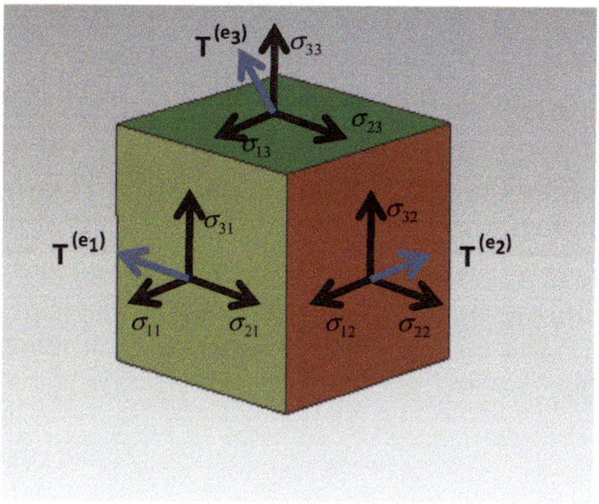

Fig. 3.1 Tensor components and their directions

than 1 are called dyad and triad which are tensors of rank 2 and 3, respectively. The dyads have a magnitude and two directions with a total of nine components, whereas triads have a magnitude and three directions resulting in 27 components. Dyads have two basis vectors for each of the nine components, whereas triads have three basis vectors for each of the 27 components.

In a tensor of any degree/order, the numerical elements are arranged as a multidimensional array with indices indicating the direction of elements. The dimensionality of the array gives the order/degree of tensor. Figure 3.1 shows the components and their directions of a tensor of rank 2.

An example of representation of a typical tensor of rank 2 is given by

$$\begin{bmatrix} \sigma_{11} & \sigma_{12} & \sigma_{13} \\ \sigma_{21} & \sigma_{22} & \sigma_{23} \\ \sigma_{31} & \sigma_{32} & \sigma_{33} \end{bmatrix}$$

The first subscript of each tensor element keeps track of the plane the component acts on (described by its unit normal vector), while the second subscript keeps track of the direction. Each component represents a magnitude for that particular plane and direction.

Let σ be a force tensor (dyad) of an object. For example, consider forces inside an object. Imagine various surfaces inside an object whose area vector is pointed to various directions. For all such surfaces, there exists a force having components pointed to any of the basis vector directions. More clearly, σ_{12} represents a force in the direction of the basis vector '1' whose surface is directed toward direction of basis vector '2.'

A second-order tensor defines an operation that transforms a vector to another vector. For example, tensor '*A*' transforms a vector '*a*' into another vector '*b*' through a dot product as

$$A : a \rightarrow b \text{ or } b = Aa \tag{3.3}$$

Any linear transformations such as rotation, reflection and projection on vectors are examples of tensors. In terms of transformation function, tensor connects vectors a and b through a linear relation

$$b = T(a) \tag{3.4}$$

The above relation obeys both homogeneity and superposition principles given by

$$T(ma) = mT(a) \tag{3.5}$$

$$T(a + b) = T(a) + T(b) \tag{3.6}$$

For any arbitrary vector 'c,' the tensor $a \otimes b$ or ab associates a vector whose direction is along 'a' and magnitude is equal to projection of vector 'c' onto 'b'. 'ab' is a tensor if and only if ab satisfies the relation.

$$(ab) \cdot c = a(b \cdot c) \tag{3.7}$$

3.3 Importance of Tensor Analysis in Electromagnetics

The theory of special relativity states that all non-gravitational laws in physics have the same form in all coordinate systems. The advent of this theory has led to the development of tensors and tensor transformation law. All the physical laws could be expressed in a concise form using tensors. The scope of tensor analysis lies in the principle of invariance, i.e., the properties of a tensor are independent of the reference frame that is used to represent the objects. Any tensor represented using a set of components in a reference frame can be related to components in a new coordinate system in an explicit way. Thus, tensor has become an ideal tool for the analysis of various laws in physics and electromagnetics.

Electromagnetics is a science that deals with electric and magnetic forces in the nature. An insight into the electromagnetic field theory reveals a thorough understanding of various discoveries and inventions in the world. The development of electromagnetics has revolutionized diverse domains such as satellite communication, radar systems and remote sensing. The diverse aerospace platforms that deal with complex structures of higher-order quadric surfaces and their hybrids have gained great momentum through the comprehensive study of field theory. Further, computational electromagnetics in aerospace domain deals with the volumetric analysis of the material properties of the composites used for the reduction of radar cross section [3]. The physical quantities whose analysis requires a track of the amplitude, direction and the components at the various planes make the EM analysis to go for tensors of rank 2 or even more.

The invisibility cloaking introduced as a breakthrough in the stealth technology in aerospace world has revealed the importance of tensor theory in electromagnetics. Electromagnetic cloaking is a phenomenon of bending the electromagnetic waves around an object without permitting absorption, reflection or scattering from that object. The cloaking can be achieved with the help of layers of periodic structures with spatially varying electromagnetic properties which can redirect the EM waves in the desired manner around the object to be hidden [4]. The properties of metamaterial, an artificially engineered structure, vary depending on the orientation or direction of excitation field [2, 1]. The coordinate transformation which is the fundamental principle behind cloaking causes the variation of property along the specified coordinate directions. The conformal antenna analysis is another major area which gives due importance to tensor analysis. In short, tensors provide a precise mathematical framework for diverse areas in field theory and therefore have an important niche in the formulation and solution of complex electromagnetic problems.

3.4 Scalars: An Electromagnetic Perspective

Scalar is a zeroth-order tensor which has only a magnitude. The scalar quantities in electromagnetic field theory are explained in the succeeding section.

3.4.1 Electric Charge as a Scalar

An atom consists of electrons, protons and neutrons. Electric or electrostatic charge is the fundamental characteristic of matter that represents how much extent it possesses more or fewer electrons than protons. The electric charge can be negative or positive based on the difference between the number of protons and electrons in it. If the number of protons is greater than number of electrons, then charge is said to be positive and vice versa. Electric charge is a scalar quantity since it has only magnitude which may be positive or negative. The charge has no direction. It is measured in coulombs. Electron is one among the fundamental constituents of an atom. Since an electron has a finite size mass and is negatively charged, group of electrons occupy a finite volume. In electromagnetics, we refer to a term ‘point charge’ which is a small but concentrated region of charged particles.

3.4.2 Electric Charge Density as a Scalar

A charge can be continuously distributed along a line or over a surface or throughout a volume. For a linear charge distribution, linear charge density gives the charge per unit length. For a sheet charge distribution, surface charge density gives the charge per unit area and volume charge density gives the charge per unit volume of a volume charge distribution.

$$\text{Linear charge density, } \rho_l = \frac{Q}{l}\ \text{C/m} \tag{3.8}$$

$$\text{Surface charge density, } \rho_s = \frac{Q}{s}\ \text{C/m}^2 \tag{3.9}$$

$$\text{Volume charge density } \rho_v = \frac{Q}{v}\ \text{C/m}^3 \tag{3.10}$$

where Q is the total charge, l is the length of the line charge distribution, s is the surface area of the sheet charge distribution, and v is the volume of the volume charge distribution [5].

3.4.3 *Electric Flux as a Scalar*

Electric flux lines are always equally spaced, and it emerges from a positive charge +Q and ends at an equal negative charge −Q or at infinity in the absence of negative charge. Assume charges +Q and −Q are present. Then, there exist electric flux tubes connecting these charges from +Q to −Q as shown in Fig. 3.2.

Flux tubes represent a constant amount of electric charge or electric flux. One coulomb electric charge produces one coulomb electric flux. Even though the

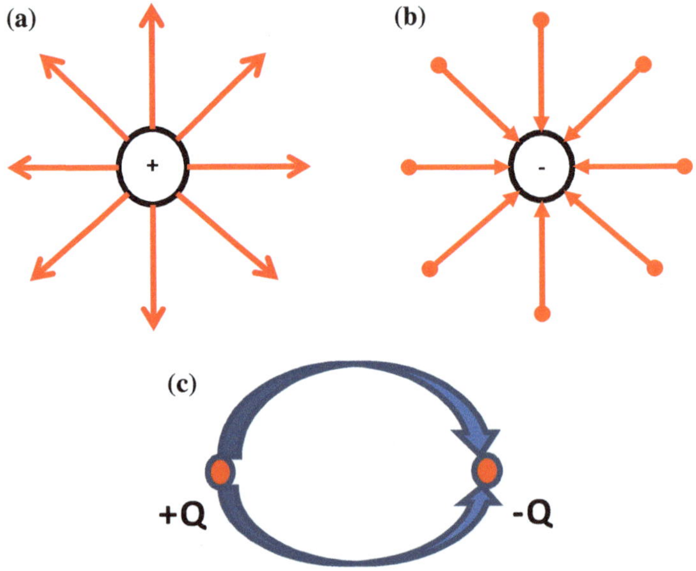

Fig. 3.2 Direction of flux lines for **a** positive charge, **b** negative charge, **c** dipole

electric lines of flux are directed as shown in Fig. 3.2, electric flux which is equal to electric charge has only magnitude and is a scalar quantity.

3.4.4 Dielectric Permittivity as a Scalar

Dielectrics are materials in which the electrons are well bound and are in a stable condition. Therefore, the ordinary electric fields do not cause any movement of charges in a dielectric and thus act as a good insulator. Permittivity can be described as a resistance experienced by a medium when excited with an electric field.

For isotropic materials, the properties of the material do not change with direction of field excitation. The molecular structure of isotropic materials is randomly oriented. The material will be polarized irrespective of the direction of field excitation. Therefore, dielectric permittivity acts as a scalar quantity in isotropic materials.

3.4.5 Electric Potential as a Scalar

The work done per unit charge to move the charge in a uniform or non-uniform electric field is termed as electric potential. Assume a charge residing at a point x_2 as shown in Fig. 3.3. Suppose charge is moving in a direction opposite to that of electric field, i.e., from x_2 to x_1, then field exerts a force on the charge and it requires work to move the charge against the force.

Electric potential is a scalar since it is the magnitude of force per unit charge times the distance through which charge is moved. A positive value for the potential indicates that work is required to move the charge, and there is a rise in potential while moving the charge in a direction opposite to that of electric field.

3.4.6 Divergence of a Vector Gives a Scalar

The divergence gives the measure of how the field is emanating from a point. It is a scalar quantity which can be positive or negative depending on the direction of

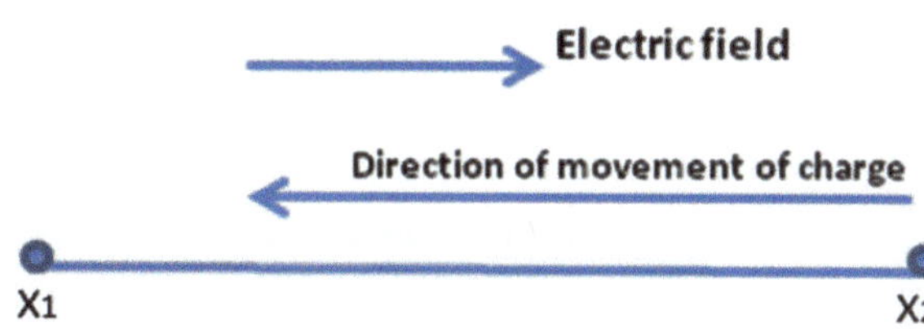

Fig. 3.3 Point charge placed in an electric field

vector field. Divergence of a vector V at a point X is the outward flux per unit volume as the volume tends to zero at that point. Since the calculation of divergence involves only scalar multiplication and addition, the divergence of a vector field results in a scalar field.

$$\text{div}A = \nabla \cdot A = \lim_{\nabla v \to 0} \frac{\oint A \cdot \mathrm{d}S}{\nabla v} \tag{3.11}$$

3.5 Vectors: An Electromagnetic Perspective

The physical quantities or entities in electromagnetics which act as a vector, i.e., having magnitude and direction, are being explained in this section.

3.5.1 *Electric Force Between Charges as a Vector*

The first electrostatic experiment conducted by Coulomb resulted in an inference that the magnitude of force between two small charged bodies Q_1 and Q_2 is proportional to product of the charges Q_1Q_2 and inversely proportional to the square of distance (r) between them [6].

$$F \propto \frac{Q_1 Q_2}{r^2} \tag{3.12}$$

The proportionality constant $k = \frac{1}{4\pi\varepsilon}$, where ε is the permittivity of the medium where charges reside. So the force between the charges can be given as a vector with magnitude and direction as

$$F = \frac{Q_1 Q_2}{4\pi\varepsilon r^2}\hat{r} \tag{3.13}$$

$\hat{r}$ is the direction of the force vector.

This force which is a vector has a direction either outward or inward on a line connecting the charges. Two like charges cause the force to be repulsive, i.e., toward outward direction, and charges of opposite signs result in an attractive force and toward inward direction as illustrated in Fig. 3.4a, b.

3.5.2 *Electric Field Intensity as a Vector*

A point charge +Q placed at a particular point will experience a radially outward force if another point charge is taken to its vicinity. Figure 3.5 shows the electric

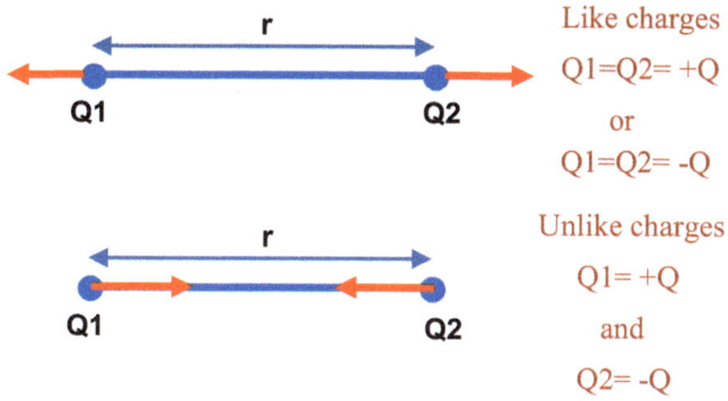

Fig. 3.4 Direction of force for **a** like charges and **b** unlike charges

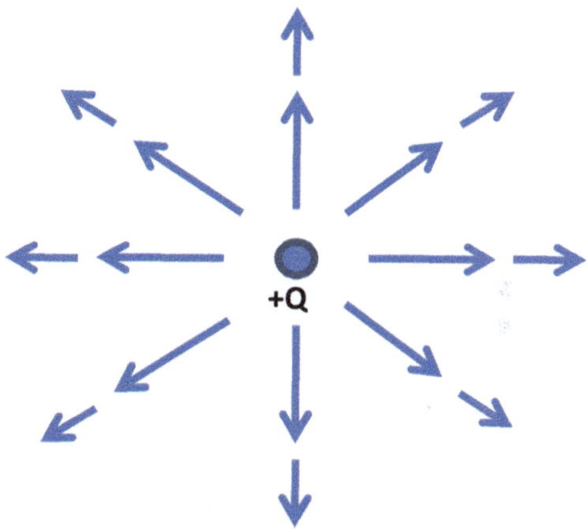

Fig. 3.5 Direction of electric field for a positive point charge

field strength around charge $+Q$. The length of vector gives the magnitude of force at the center of the vector. Force per unit charge is defined as the electric field intensity. This electric field intensity vector has the same direction as that of force but a different magnitude.

$$E = \frac{F}{Q} = \frac{Q}{4\pi\varepsilon r^2}\hat{r} \tag{3.14}$$

3.5.3 *Electric Dipole Moment as a Vector*

Two equal but opposite charges separated by a small distance constitute an electric dipole. As shown in Figure, the charges $+Q$ and $-Q$ placed l distance apart will produce an electric dipole moment which is a vector having a magnitude of Q_l direction pointing from $-Q$ to $+Q$.

3.5.4 Electric Flux Density as a Vector

If there are point charges $+Q$ and $-Q$, then there exist electric flux lines or tubes connecting these charges from $+Q$ to $-Q$. Flux tubes as shown in Fig. 3.6 represent a constant amount of electric charge or electric flux. The electric flux density is the electric flux or charge per unit cross-sectional area of the flux tube.

$$D = \frac{\psi}{A}\ \mathrm{C/m^2} \tag{3.15}$$

Imagine a point charge $+Q$ uniformly distributed over a sphere of radius r as shown in Fig. 3.7. Then, the electric flux density at a distance r from the center is given by

$$D = \frac{Q}{4\pi r^2}\ \mathrm{C/m^2} \tag{3.16}$$

where $4\pi r^2$ is the surface area of the sphere.

The electric field strength around an isolated point charge Q is given by

$$E = \frac{Q}{4\pi\varepsilon_0 r^2}\hat{r} \tag{3.17}$$

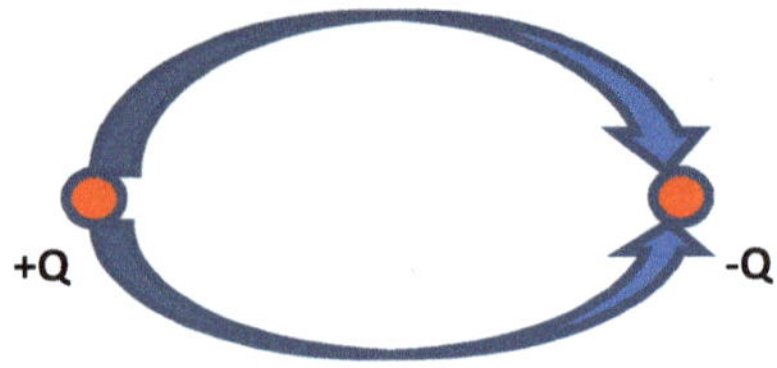

Fig. 3.6 Direction of flux lines (tubes) in a dipole

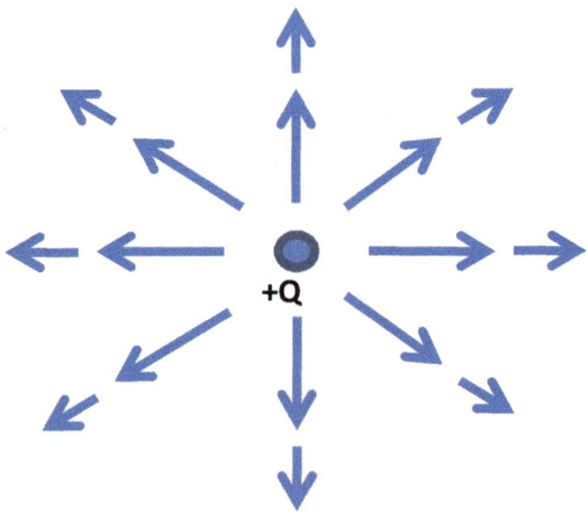

Fig. 3.7 Direction of flux lines for a charge +Q

$$\varepsilon_0 E = \frac{Q}{4\pi r^2}\hat{r} = D \quad (3.18)$$

$$D = \varepsilon_0 E \quad (3.19)$$

Therefore, the electric field intensity when multiplied by permittivity gives charge per unit area, i.e, the electric flux density. Therefore, electric flux density is a vector whose magnitude is proportional to charge and direction same as that of electric field. For any isotropic media whose properties do not change with direction, the flux density and electric field vectors have the same direction.

3.5.5 *Gradient of a Scalar as a Vector*

Gradient of a scalar field is a vector which has a magnitude of maximum spatial rate of change of the field per unit distance and a direction of the same. For example, in Cartesian coordinates, the gradient of a scalar can be expressed as

$$\nabla V = \frac{\partial v}{\partial x} a_x + \frac{\partial v}{\partial y} a_y + \frac{\partial v}{\partial z} a_z \quad (3.20)$$

An example of a vector field in electromagnetics which can be given as the gradient of a scalar is explained below. Consider an electric field which is directed from point X_1 to point X_2. Let a charge be moving from X_2 to X_1, against the direction of field, then work is required to move the charge and there occurs a rise in potential as illustrated in Fig. 3.8. The gradient of scalar electric potential means the measure of rise or fall in potential. The gradient of potential is a vector having magnitude equal to electric field intensity and direction opposite to that of electric field.

$$\nabla V = -E \quad (3.21)$$

The direction of gradient of potential at any point is normal to the constant potential or equipotential surface.

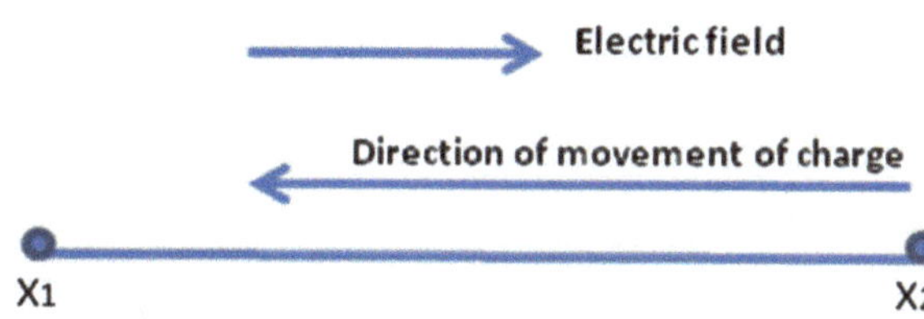

Fig. 3.8 Point charges placed in an electric field

3.5.6 *Curl of a Vector as a Rotational Vector*

The curl is a measure of circulation of a vector field. This rotational vector has a magnitude equal to maximum rotation of A per unit area as area shrinks to zero and a direction normal to the area when the area is oriented to make a maximum rotation. Therefore, curl of a vector is another vector. Mathematically, curl of a vector field A can be expressed as

$$\mathrm{curl}A = \nabla \times A = \left(\lim_{\nabla S \to 0} \frac{\oint A \cdot \mathrm{d}l}{\nabla S}\right)_{\max} \hat{a}_n \tag{3.22}$$

The physical significance of curl reveals the extent of rotation of a vector field around a point. For a field aligned along straight line, the curl is zero.

3.5.7 *Magnetic Field or Flux Lines Due to a Current-Carrying Wire as a Vector*

Any current-carrying wire will have a magnetic field line forming closed loops around it as depicted in Fig. 3.9. This is in accordance with the right-hand thumb rule which states that if current is in the direction of thumb, the magnetic field will be in the direction of fingers forming a closed loop around the wire.

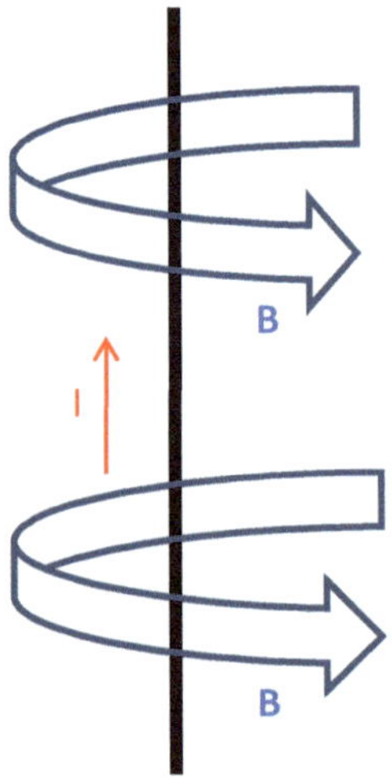

Fig. 3.9 Magnetic flux lines in a current-carrying wire

3.5.8 *Magnetic Flux Density as a Vector*

Magnetic flux density at a point is a vector having direction same as that of magnetic field lines at that point. Consider a wire carrying current *I*. The magnetic flux density at a point on the right of the wire is in an inward direction normal to the wire. Similarly, magnetic flux density at a point on the left of the wire is in an outward direction normal to the wire.

(a) *Infinite straight conductor*

Figure 3.10 illustrates the parameters related to the calculation of flux density of an infinite current-carrying wire. According to Biot–Savart law, the magnetic flux density at a point *P* is given by

$$\mathrm{d}B = \frac{\mu}{4\pi}\frac{Idl\sin\theta}{r^2} \tag{3.23}$$

Integrating the above equation, the total flux density is obtained as

$$B = \frac{\mu I}{4\pi}\int\frac{\sin\theta}{r^2}\mathrm{d}l \tag{3.24}$$

Simplifying, we get

$$B = \frac{\mu I}{2\pi R} \tag{3.25}$$

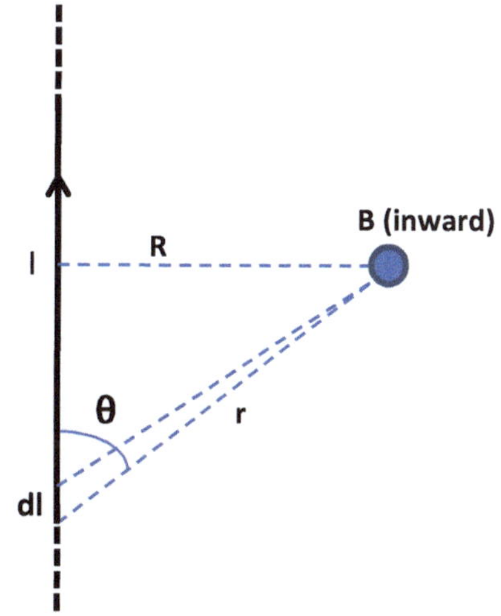

Fig. 3.10 Magnetic flux density for a straight current-carrying wire

(b) *Current-carrying loop*

The magnetic flux density at a point on the axis of a loop of radius, R, carrying current I is a vector whose direction is normal to the radius vector joining the current element dl to the point on the loop axis as shown in Fig. 3.11.

The flux density vector can be resolved into two components, one along the axis of the loop dB_z and other component dB_n perpendicular to the axis. Integrating dB_n for all elements around the loop results in cancelation of all components due to symmetry. Therefore, only B_z exists.

$$\mathrm{d}B_z = \frac{\mu I R^2}{4\pi (R^2 + z^2)^{3/2}} \mathrm{d}\phi \tag{3.26}$$

$$B_z = \frac{\mu I R^2}{4\pi (R^2 + z^2)^{3/2}} \int_0^{2\pi} \mathrm{d}\phi = \frac{\mu I R^2}{2(R^2 + z^2)^{3/2}} \tag{3.27}$$

3.5.9 *Force Due to a Current-Carrying Wire as a Vector*

Assume a wire carrying current I is placed in a magnetic field of flux density B as illustrated in Fig. 3.12. The magnitude of force experienced on the wire will be

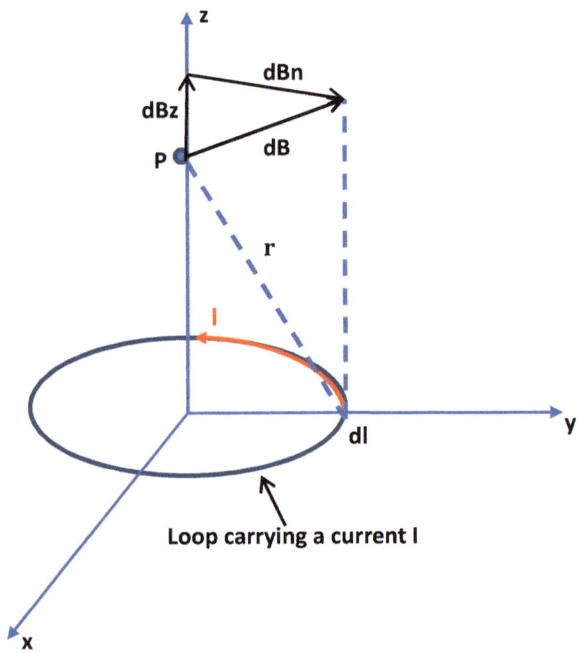

Fig. 3.11 Magnetic flux density for a current-carrying loop

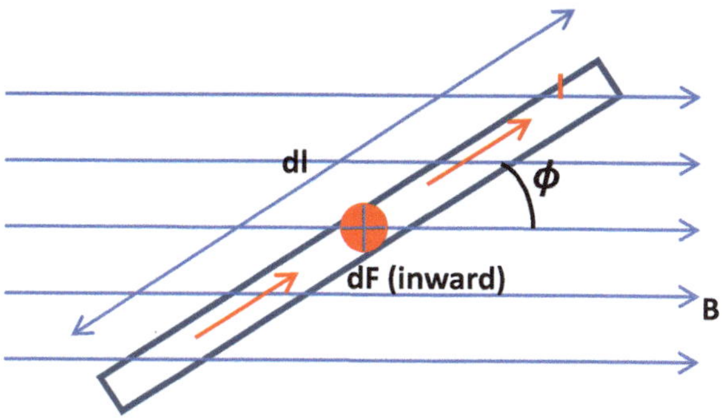

Fig. 3.12 Force in a current-carrying wire placed in magnetic field

$$F = BIL\sin\phi \tag{3.28}$$

where L is the length of the wire placed in magnetic field. The direction of this force can be explained by modifying this equation using a vector form as

$$F = (I \times B)L \tag{3.29}$$

Force is expressed as a cross-product of current and flux density vectors. Therefore, force is in a direction normal to the plane containing the current-carrying element (i.e., current) and magnetic flux density B. The force due to an infinitesimal current-carrying element dl is given by

$$\mathrm{d}F = BIdl\sin\phi \tag{3.30}$$

$$\mathrm{d}F = (I \times B)\,dl \tag{3.31}$$

where ϕ is the angle made by the current element with the magnetic field.

For a point charge Q moving with a velocity v, the force on the charge due to magnetic field is given by

$$F = Q(v \times B) \tag{3.32}$$

(since $IL = (Q/t)\,L = Qv$).

Therefore, force vector is in a direction perpendicular to magnetic flux density and direction of movement of point charge.

3.5.10 Force Between Two Parallel Conductors Carrying Current as a Vector

Two parallel linear conductors of length l are placed at a distance R apart. Assume conductor 1 carries a current I and conductor 2 carries I'. Then, magnitude of force on conductor 2 due to conductor 1 is given by

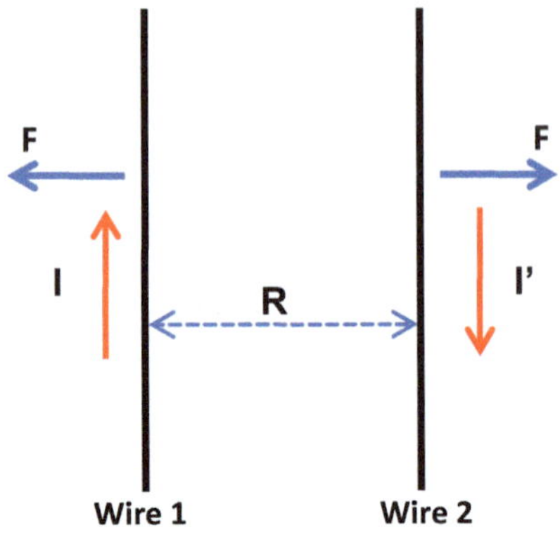

Fig. 3.13 Force between two current-carrying wires placed in a magnetic field

$$F = BI'l \tag{3.33}$$

Substituting the magnetic flux of conductor 1, F can be obtained as

$$\begin{aligned} B &= \frac{\mu I}{2\pi R} \\ F &= \frac{\mu I}{2\pi R} I'l \end{aligned} \tag{3.34}$$

Assuming conductor 1 carrying current in a direction opposite to that of conductor 2, the force will be outward and conductors repel each other as depicted in Fig. 3.13.

When both conductors are carrying current in the same direction, forces will be inward and conductors will be attracted.

3.5.11 Torque and Magnetic Moment as Vectors

For a loop carrying a current *I*, placed in a magnetic field of uniform flux density *B* parallel to the plane of the loop, product of tangential force on the loop and radial distance where the force acts gives the torque. The loop tends to rotate due to this torque.

The torque produced in the loop can be given by

$$T = BILd \sin\phi \tag{3.35}$$

Since area of the loop $A = Ld$,

$$T = BIA \sin\phi \tag{3.36}$$

Here, $IA = m$, the magnetic moment whose direction is perpendicular to the plane of the loop (right-hand rule; fingers in the direction of current and thumb in the direction of magnetic moment). Therefore, the torque can be modified as

$$T = mB\sin\phi = m \times B \tag{3.37}$$

Therefore, torque vector has a direction coinciding with the axis of rotation of the loop.

3.5.12 Magnetic Field Intensity Vector

For any isotropic media, magnetic field is given as the ratio of magnetic flux density to dielectric permeability of the media.

$$H = \frac{B}{\mu} \tag{3.38}$$

Therefore, the magnetic field has the same direction as that of the magnetic flux density.

3.5.13 Magnetic Vector Potential

According to Maxwell's equation, the divergence of magnetic flux density is zero. Therefore, the magnetic flux density can be given as the curl of another vector.

$$B = \nabla \times A \tag{3.39}$$

where A is known as the magnetic vector potential. Unlike scalar electric potential, the magnetic potential is a vector potential function. For any current distribution, the magnetic potential can be given as

$$A = \frac{\mu}{4\pi}\iiint_{v} \frac{J}{r}\mathrm{d}v \tag{3.40}$$

where J is the current density of elemental volume and r is the distance from each element to the point where the magnetic potential is calculated. Though magnetic potential does not have much physical significance, it can be used to evaluate the magnetic field easily which is indispensable in most of the complex electromagnetic problems.

3.6 Tensors: An Electromagnetic Perspective

The tensors of ranks greater than 1 are discussed in detail from a mathematical as well as electromagnetic perspective in this section.

3.6.1 Dyads: Tensors of Rank 2

A dyad is a second-order tensor given as an outer product of two vectors

$$A = a \otimes b \tag{3.41}$$

Dyadic is a linear combination of dyads with scalar coefficients

$$B = (a \otimes b) + (c \otimes d) \tag{3.42}$$

Properties

$$(a \otimes b)(\alpha c + d) = \alpha (a \otimes b) c + (a \otimes b) d \tag{3.43}$$

$$(\alpha a + \beta b) \otimes c = \alpha (a \otimes c) + \beta (b \otimes c) \tag{3.44}$$

$$A(a \otimes b) = (Aa) \otimes b \tag{3.45}$$

Determinant of tensor

The determinant of a tensor can be described in a similar approach used in matrix algebra [7]. The determinant of tensor A is given by

$$\det A = \det \begin{bmatrix} A_{11} & A_{12} & A_{13} \\ A_{21} & A_{22} & A_{23} \\ A_{31} & A_{32} & A_{33} \end{bmatrix} \tag{3.46}$$

Properties

$$\det (AB) = \det A \det B \tag{3.47}$$

$$\det (A^T) = \det A \tag{3.48}$$

Inverse of tensor

The inverse of a tensor can be explained using matrix algebra. The inverse of a tensor A is represented by A^{-1}

$$AA^{-1} = I = A^{-1}A \tag{3.49}$$

where I is the identity tensor matrix

Properties

$$(AB)^{-1} = B^{-1}A^{-1} \tag{3.50}$$

$$\det(A^{-1}) = (\det A)^{-1} \tag{3.51}$$

$$(A^{-1})^T = (A^T)^{-1} \tag{3.52}$$

$$(\alpha A)^{-1} = \frac{1}{\alpha}A^{-1} \tag{3.53}$$

$$(A^{-1})^{-1} = A \tag{3.54}$$

Orthogonality of tensor

Tensor 'A' is said to be orthogonal if

$$AA^T = I \tag{3.55}$$

If $\det A = 1$, then A is a proper orthogonal tensor and A corresponds to a rotation.

If $\det A = -1$, then A is an improper orthogonal tensor and A corresponds to a reflection.

Covariance and contra-variance of tensors

For a vector, as the basis changes, the components of the vector also change. Similarly, tensor components also change according to the coordinate transformation or change of basis. If the tensor components transform in a similar manner with a basis transformation matrix, then it is said to be a covariant, and if the tensor components transform with the inverse of the basis transformation, then it is said to be contravariant. The covariant tensor is denoted using a subscript (e.g., F_{12}) and contravariant using a superscript (e.g., F^{12}).

Metric tensor

Consider a vector x having components x^i in a set of orthogonal coordinate axes. The magnitude of the vector x can be given as

$$|x| = \sqrt{x^i x^i} \tag{3.56}$$

Consider two adjacent points Q and Q' separated by a distance 'dx.' The square of the distance between Q and Q' can be given as

$$ds^2 = dx^i dx^i \tag{3.57}$$

Change in coordinate system described by the given coordinate transformation $x^i = x^i(y^1, \ldots y^n)$ will modify the square of the distance between Q and Q' in the new reference frame as

$$ds^2 = \frac{\partial x^i}{\partial y^\alpha}\frac{\partial x^i}{\partial y^\beta} dy^\alpha dy^\beta \tag{3.58}$$

$$ds^2 = g_{\alpha\beta}\, dy^\alpha dy^\beta \tag{3.59}$$

where $g_{\alpha\beta}(y) = \frac{\partial x^i}{\partial y^\alpha}\frac{\partial x^i}{\partial y^\beta}$ is the metric tensor whose components are functions of y.

The metric tensor is symmetric with respect to the indices α and β. All the metric properties of space can be determined using this metric tensor.

3.6.2 *Dielectric Permittivity in Anisotropic Media as a Tensor*

For any isotropic media, the properties of the material do not vary with respect to the direction. The molecules of an isotropic material will be polarized independent of the direction of electric field. In isotopic media, the relation between electric flux density D and electric field can be given as

$$D = \varepsilon E \tag{3.60}$$

where ε is a scalar constant known as the dielectric permittivity of the medium.

In anisotropic materials, the polarization of molecules will depend on the direction of electric field. There may be strong polarization along one direction of electric field, whereas the molecules need not be polarized on application of field along the other directions. Figures 3.14 and 3.15 show the way in which the anisotropic material is polarized according to the direction of electric field excitation.

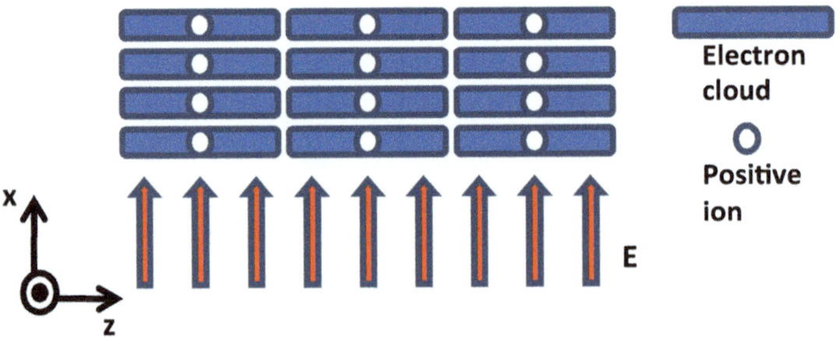

Fig. 3.14 Anisotropic material unpolarized with electric field along x direction

Fig. 3.15 Anisotropic material strongly polarized with electric field along z direction

The effect of polarization in an anisotropic material can be well explained using tensors. In anisotropic media, the electric flux density D is related to electric field given by a dielectric permittivity tensor matrix.

$$\begin{bmatrix} D_x \\ D_y \\ D_z \end{bmatrix} = \begin{bmatrix} \varepsilon_{xx} & \varepsilon_{xy} & \varepsilon_{xz} \\ \varepsilon_{yx} & \varepsilon_{yy} & \varepsilon_{yz} \\ \varepsilon_{zx} & \varepsilon_{zy} & \varepsilon_{zz} \end{bmatrix} \begin{bmatrix} E_x \\ E_y \\ E_z \end{bmatrix} \tag{3.61}$$

The x, y and z components of electric flux density are related to x, y and z components of electric field by a permittivity matrix. The value of the dielectric permittivity varies depending upon the direction of excitation field.

Dielectric tensor matrix is Hermitian, i.e., $\varepsilon_{ij} = \varepsilon_{ji}^*$.

For any lossless material, the elements of permittivity tensor matrix are real. Therefore, tensor matrix is symmetric, i.e., $\varepsilon_{ij} = \varepsilon_{ji}$, and the tensor can be represented using six elements instead of nine.

$$\varepsilon_{ij} = \begin{bmatrix} \varepsilon_{11} & \varepsilon_{12} & \varepsilon_{13} \\ \varepsilon_{12} & \varepsilon_{22} & \varepsilon_{23} \\ \varepsilon_{13} & \varepsilon_{23} & \varepsilon_{33} \end{bmatrix} \tag{3.62}$$

since $\varepsilon_{12} = \varepsilon_{21}$, $\varepsilon_{13} = \varepsilon_{31}$, $\varepsilon_{23} = \varepsilon_{32}$.

3.6.3 *Dielectric Permeability in Anisotropic Media as a Tensor*

The relation between magnetic flux density and the magnetization is given by

$$B = \mu H \tag{3.63}$$

where μ is the permeability of the medium.

For free space, the dielectric permeability is a scalar whose magnitude $\mu_o = 4\pi \times 10^{-7}$. Since μ is a scalar, the flux density and the magnetization have same direction but different magnitude. In anisotropic materials (e.g., non-cubic crystals), the component atoms have some peculiar dipole properties that make these terms differ in both magnitude and direction. In such materials, the scalar permeability is then replaced by the tensor permeability μ. The permeability becomes a tensor of rank 2. B and H are both vectors, but they now differ from one another in both magnitude and direction.

3.6.4 Electric Susceptibility as a Tensor

Electric susceptibility gives the measure of how well (polarized) the material responds to electric field. In any isotropic material electric polarization, vector is related to electric field vector as

$$\overrightarrow{P} = \chi\overrightarrow{E} \tag{3.64}$$

where χ is the dielectric susceptibility which is a scalar constant. Here $\overrightarrow{P}$ and $\overrightarrow{E}$ are parallel vectors.

In any material, charge acts as the source of polarization when electric field excitation is applied. In anisotropic materials, the electrostatic potential where the charges reside is not symmetric. Therefore, the material polarization is dependent on the orientation of charges inside the media. In anisotropic materials, electric polarization vector is related to electric field vector using χ which is the susceptibility tensor, i.e., the value of susceptibility varies with direction of excitation field. Therefore, $\overrightarrow{P}$ and $\overrightarrow{E}$ may not be parallel vectors. Polarization may not be in the same direction as driving field. Figure 3.16a, b shows a comparison between the effect of polarization in isotropic and anisotropic material.

The electric polarization vector can be given as

$$P_i = \sum_{j=1}^{3} \chi_{ij} E_j \tag{3.65}$$

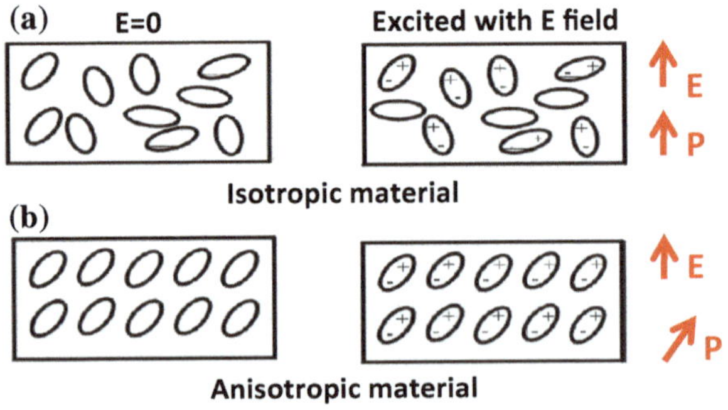

Fig. 3.16 Polarization in **a** isotropic material and **b** anisotropic material

where $i = 1, 2, 3$

$$\vec{P} = \begin{bmatrix} \chi_{11} & \chi_{12} & \chi_{13} \\ \chi_{21} & \chi_{22} & \chi_{23} \\ \chi_{31} & \chi_{32} & \chi_{33} \end{bmatrix} \vec{E} \tag{3.66}$$

or can be given as

$$\begin{aligned} P_1 &= \chi_{11}E_1 + \chi_{12}E_2 + \chi_{13}E_3 \\ P_2 &= \chi_{21}E_1 + \chi_{22}E_2 + \chi_{23}E_3 \\ P_1 &= \chi_{31}E_1 + \chi_{32}E_2 + \chi_{33}E_3 \end{aligned}$$

3.6.5 Electromagnetic Field Tensor

The electric fields can be transformed to magnetic fields and vice versa. It is also termed as Maxwell bi-vector or Faraday tensor or simply field strength tensor. For a stationary point charge, electric field vector exists. But for a moving charge, electric and magnetic fields coexist which result in an electromagnetic field which can be represented using a tensor.

$$F_{\mu\nu} = \begin{bmatrix} 0 & iE_x & iE_y & iE_z \\ -iE_x & 0 & B_z & -B_y \\ -iE_y & -B_z & 0 & B_x \\ -iE_z & B_y & -B_x & 0 \end{bmatrix} \tag{3.67}$$

The electromagnetic tensor is composed of six independent fields (electric and magnetic). For example, in Cartesian coordinates, the tensor elements are formed out of E_x, E_y, E_z, B_x, B_y and B_z components.

Electromagnetic tensor is also called Maxwell bi-vector because of its anti-symmetry property

$$F_{\mu\nu} = -F_{\nu\mu} \tag{3.68}$$

The electromagnetic tensor can also be described using vector potential as

$$F_{\mu\nu} = \frac{\partial A_\nu}{\partial x_\mu} - \frac{\partial A_\mu}{\partial x_\nu} \tag{3.69}$$

For example,

$$F_{12} = \frac{\partial A_2}{\partial x_1} - \frac{\partial A_1}{\partial x_2} = (\nabla \times A)_z = B_z \tag{3.70}$$

which is the element in the first row, second column (starting from top leftmost element as zeroth row and zeroth column). Similarly,

$$F_{13} = \frac{\partial A_3}{\partial x_1} - \frac{\partial A_1}{\partial x_3} = -(\nabla \times A)_y = -B_y \tag{3.71}$$

Thus, mapping of each element in tensor matrix can be proved. Electromagnetic tensor does not change with respect to reference frame as electric and magnetic fields vary. Therefore, electromagnetic tensor is said to be isomorphic.

3.7 Conclusion

The scalars, vectors and tensors have been thoroughly investigated from an electromagnetic perspective for their application in aerospace platforms. The properties according to their nature can be classified based on the parameters used to represent it. These possess extensive applications in the formulation and solution of complex engineering problems in aerospace domain.

References

1. Chen, F., X. Wang, and E. Semouchkina, "*Resonance mode splitting in split-ring resonator arrays used in the microwave invisibility cloak*," Proceedings of Antennas and Propagation Society International Symposium (APSURSI), 2012.
2. Schurig, D., J. J. Mock, B. J. Justice, S. A. Cummer, J. B. Pendry, A. F. Starr, and D. R. Smith, "*Metamaterial electromagnetic cloak at microwave frequencies*," Science Express, vol. 314, pp. 977–979, 2006.
3. Ahmad B. N. and M. S. Sharawi, "*Extraction of material parameters for metamaterials using a full-wave simulator*," IEEE Antennas and Propagation Magazine, vol. 55, no. 5, pp. 202–211, Oct. 2013.
4. Ahn, D., "*Calculation of permittivity tensors for invisibility devices by effective medium approach in general relativity*," Document Submitted to Center for Quantum Information Processing, Department of Electrical and Computer Engineering, University of Seoul, Seoul 130–743, Republic of Korea, pp. 1–40, 2006.
5. Krauss, J. D., and Carver, K. R., *Electromagnetics*, McGraw-Hill, Kogakusha, ISBN 0-07-035396-4, 1973.
6. Sadikku, M. N. O., *Principles of Electromagnetics*, Oxford University Press, ISBN 978-0-19-806229-5, 2007.
7. Sokolnikoff, I. S., *Tensor analysis Theory and Applications*, John Wiley and Sons Inc., USA, 1951.

Chapter 4
Design and Optimization of Multilayer Ideal Cloak

M.H. Jyothi and Balamati Choudhury

4.1 Introduction

Invisibility cloaking has been a dream of everybody and was only reflected in movies and stories. After introduction of metamaterial science and technology researchers put their full effort to make their dream come true and started demonstrating in a small scale. An ideal invisibility cloak is a volumetric shell that covers an object to make it invisible by bending the incident ray around the object, as shown in the Fig 4.1. The concept of invisibility cloaking has attracted many researchers especially in aerospace domain because of its exiting property of making an object ideally invisible to the observer. This invisibility cloaking concept can be formulated through optical transformation theory. According to which the volumetric free space is converted into volumetric shell where in incident waves are controlled and manipulated.

Among all the related research, cloak with spherical and circular cylindrical shell geometrics has gained much attention because, objects of any shape can be hidden in these two kinds of cloaking shapes [1]. In aerospace platform, cloaking represents one type of stealth technology that will make the electromagnetic (EM) scatterer invisible to the EM detecting devices like Radar, Sonar etc.

The term invisibility in aerospace refers to hiding an object from detection techniques like radar. Schurig et al. (2006) proposed the novel concept of cloaking through transparent gradient-index structure, in which they designed multilayer metamaterial to obtain the electromagnetic invisibility [2]. The design of the invisibility cloak is the reverse process, where the electromagnetic wave propagation direction will be known and the material properties will be obtained according to that.

M.H. Jyothi (✉) · B. Choudhury
Centre for Electromagnetics, CSIR-National Aerospace Laboratories, Old Airport Road, Bengaluru 560017, India
e-mail: jyothiece.komal@gmail.com

B. Choudhury (ed.), *Metamaterial Inspired Electromagnetic Applications*,
DOI 10.1007/978-981-10-3836-5_4

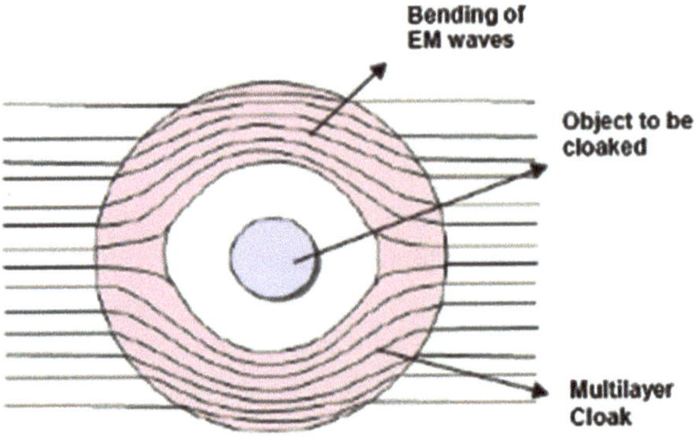

Fig. 4.1 Illusion of cylindrical cloak

Pendry et al. (2006) proposed the theory of coordinate transformation for electromagnetic fields in an anisotropic medium to calculate the redirection of the field [3]. The bending of electromagnetic wave trajectory around the cloaked object is well explained through coordinate transformation theory, which causes spatial variation in the constitutive parameters like relative permittivity and relative permeability of the electromagnetic structure.

The concept of cloaking an object from electromagnetic field is gaining momentum, as it leads to advancement in controlling and manipulating the electromagnetic radiation. Maxwell's equations are form-invariant to coordinate transformations [4]. Hence the permittivity tensor and the permeability tensor components are affected by the transformation and become both spatially variant and anisotropic. By applying these complex material properties to the cloaking shell the electromagnetic waves can be controlled and guided according to the structure without any reflection and absorption of incident EM waves. The design and analysis of cloaking shells have been carried out using finite element method (FEM)-based COMSOL Multiphysics simulation software. This chapter focuses on design and simulation of circular cylindrical and spherical cloaking shells along with the radar cross section (RCS) analysis of multilayer cloaking shells. Further, this chapter includes the idea of coordinate transformation toward the RCS reduction of spherical cloaking shell in C band (4–8 GHz) frequency range.

4.2 Design of Ideal Circular Cylindrical Cloak

Cloaking with spherical and circular cylindrical geometric shell has received much attention among all the related research. Circular cylindrical cloak design relates the cylindrical coordinate system (r, θ, z) with the Cartesian coordinate system (x, y, z) as

$$x = r\cos\theta \quad y = r\sin\theta \quad z = z \tag{4.1}$$

where r is the radius of the cylinder, θ is the angle which varies from 0° to 360°, and z is the finite height of the cylinder.

Here cylindrical region is compressed into the cylindrical shell in order to make an object inside the shell invisible. For the design of circular cylindrical cloak, let a and b be the radius of inner cylinder and outer cylinder respectively. To create a cloak, we specify a spatial transformation that maps a cylindrical region $0 < r < b$ in original coordinate (r, θ, z) into an annular region $a < r' < b$ in a new coordinate system (r', θ', z') as shown in Eq. (4.2)

$$r = a + r' * \frac{(b-a)}{b} \quad \theta = \theta' \quad z = z' \tag{4.2}$$

Using this transformation, elements of permittivity and permeability tensors [5] can be determined by solving the determinant of the following equation

$$\varepsilon^{ij} = \begin{bmatrix} \left(\frac{b-a}{b}\right) * a & 0 & 0 \\ 0 & \left(\frac{b}{b-a}\right) * \frac{1}{r'} & 0 \\ 0 & 0 & \left(\frac{b}{b-a}\right) * r' \end{bmatrix} * \frac{1}{r} \tag{4.3}$$

Determinant of the above equation is given by

$$\det\left(\varepsilon^{ij}\right) = \left(\frac{b-a}{b}\right) * r' * \left(\frac{b}{b-a}\right) * \frac{1}{r'} * \left(\frac{b}{b-a}\right) * r' * \frac{1}{r} \tag{4.4}$$

Simplified form of Eq. (4.4) is

$$\det\left(\varepsilon^{ij}\right) = \left(\frac{b}{b-a}\right)^2 * \left(\frac{r-a}{r}\right) \tag{4.5}$$

This transformation leads to the expression of permittivity and permeability tensors components of the circular cylindrical cloak.

$$\varepsilon_r = \mu_r = \left(\frac{r-a}{r}\right) \quad \varepsilon_\theta = \mu_\theta = \left(\frac{r}{r-a}\right) \quad \varepsilon_z = \mu_z = \left(\frac{b}{b-a}\right)^2 * \left(\frac{r-a}{r}\right) \tag{4.6}$$

The transformation media that forms the cloak has been laid between two cylinders. Figure 4.2 depicts the perfect electric conductor (PEC) cylinder which is to be cloaked has the radius of a = 3 cm and the cloaking cylinder surrounding the PEC has the radius of b = 6 cm with the material properties as mentioned in Eq. (4.6). Air gap is provided between PEC cylinder and cloaking shell. By applying these complex material properties, the concealed volume and the cloak appear to have the properties of free space when viewed externally.

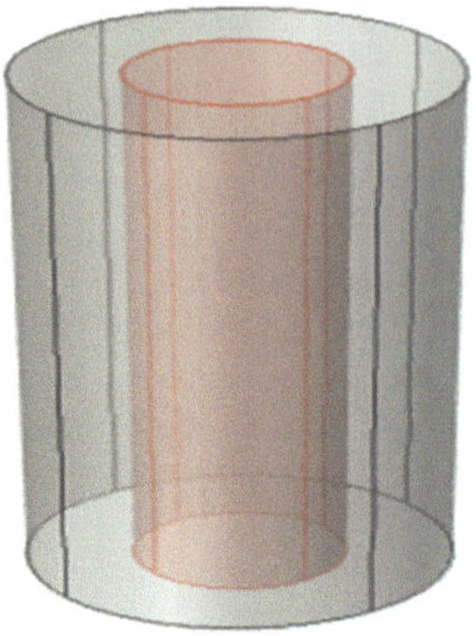

Fig. 4.2 Typical geometry of a circular cylindrical cloak

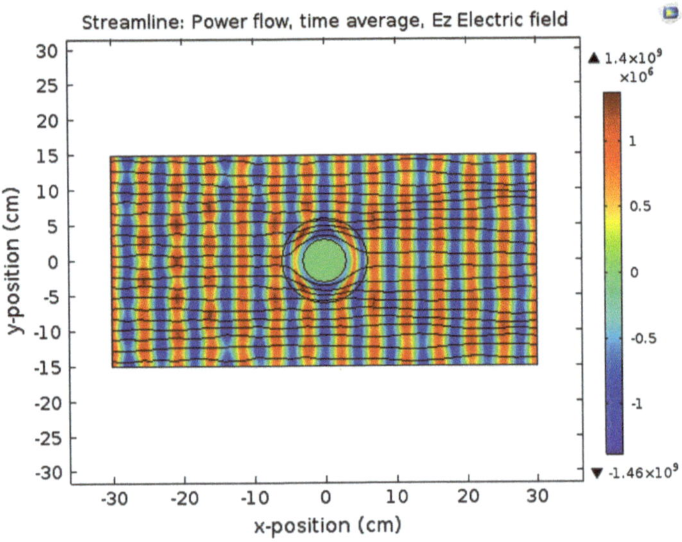

Fig. 4.3 2D cylindrical cloak, electric field patterns, with stream lines indicating the direction of power flow

The circular cylindrical cloak performance is validated by simulating it using COMSOL Multiphysics software. This approach describes the propagation of electromagnetic waves based on numerical integration of a set of Hamilton's equations obtained by taking the geometric limit of Maxwell's equations in anisotropic, inhomogeneous medium.

Figure 4.3 shows the electric field distribution of a circular cylindrical cloak obtained by applying excitation port at left side, receiving port at right side and the remaining sides are applied with periodic boundary condition of the rectangular air box. RF module of COMSOL Multiphysics software has been used for simulation. It can be seen that the waves path has been smoothly bent around the cloaked object and the waves emerges through the cloaking structure without any distortion.

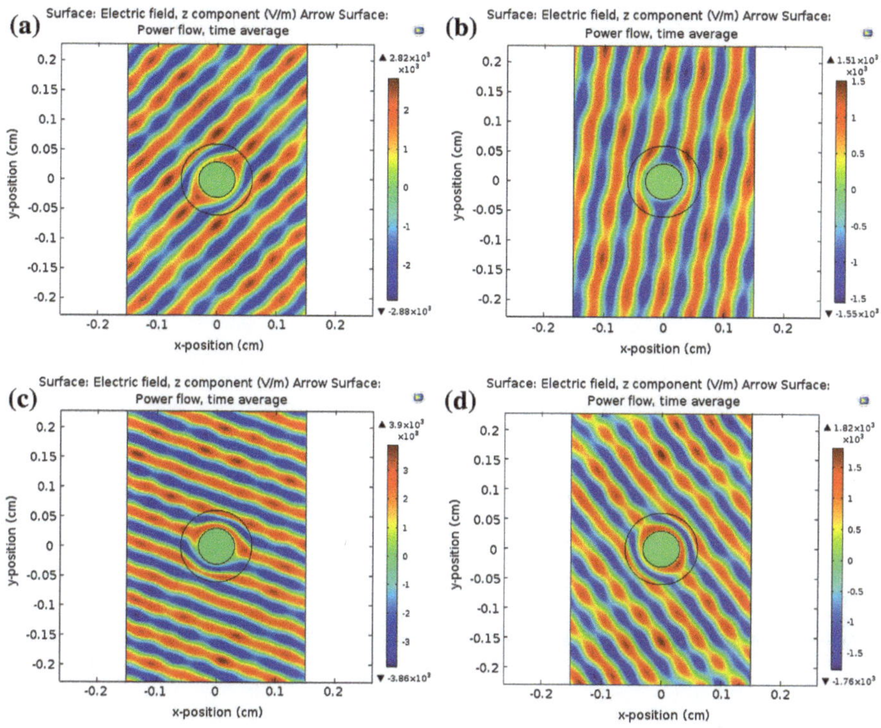

Fig. 4.4 2D circular cylindrical cloak with EM wave incident at **a** 45°, **b** 80°, **c** 200°, **d** 310°

Further to study the cloaking effect at different angle of incidence, the wave optics module of COMSOL Multiphysics software has been used. Here the transmitting port and receiving port are applied at the top and bottom of the rectangular air box used for simulation respectively and remaining sides are applied with periodic boundary condition. The electromagnetic waves incident at different oblique angles is shown in Fig. 4.4. From Figs. 4.3 and 4.4, it is clearly observed that incident rays are bending around the object and emerging out without any absorption or scattering at different angle of incidence. Thus, the developed mathematical formulations validate the bending of electromagnetic wave around the object for all the angle of incidence and become the standard basis towards the realization of cloaking effect.

4.3 Design of Ideal Spherical Cloak

The spherical cloak design relates the spherical coordinate system (r, θ, φ) with the Cartesian coordinate system (x, y, z) as

$$x = r * \sin\theta * \cos\varphi \quad y = r * \sin\theta * \sin\varphi \quad z = r * \cos\theta \tag{4.7}$$

where r is the radius of the sphere, θ is the azimuth angle varying from 0° to 180°, and φ is the elevation angle that varies from 0° to 360°.

Transformation optic theory is used to compress the spherical region into spherical shell of inner radius a and outer radius b. To create a cloak, we define a spatial transformation that maps a spherical region $0 < r < b$ in the original coordinate (r, θ, φ) into an annular region $a < r' < b$ in a new coordinate (r', θ', ϕ') as

$$r = a + r' * \left(\frac{b-a}{b}\right) \quad \theta = \theta' \quad \phi = \phi' \tag{4.8}$$

Using this transformation, the elements of the permittivity and permeability tensors [5] can be determined by solving the determinant of the following equation

$$\varepsilon^{ij} = \begin{pmatrix} \left(\frac{b-a}{b}\right) * r'^2 * \sin\theta' & 0 & 0 \\ 0 & \left(\frac{b}{b-a}\right) * \sin\theta' & 0 \\ 0 & 0 & \left(\frac{b}{b-a}\right) * \frac{1}{\sin\theta'} \end{pmatrix} * \frac{1}{r^2 * \sin\theta} \tag{4.9}$$

Determinant of the above equation is calculated as follows

$$\det(\varepsilon^{ij}) = \left(\frac{b-a}{b}\right) * r'^2 * \sin\theta' * \left(\frac{b}{b-a}\right) * \sin\theta' * \left(\frac{b}{b-a}\right) * \frac{1}{\sin\theta'} * \frac{1}{r^2 * \sin\theta} \tag{4.10}$$

Simplified form of Eq. (4.10) is

$$\det(\varepsilon^{ij}) = \left(\frac{b-a}{b}\right)^3 * \left(\frac{r-a}{r}\right)^2 \tag{4.11}$$

This transformation leads to the expression of permittivity and permeability tensors components of the spherical cloak.

$$\varepsilon_r = \mu_r = \left(\frac{b-a}{b}\right) \quad \varepsilon_\theta = \mu_\theta = \left(\frac{b-a}{b}\right) * \left(\frac{r-a}{r}\right) \quad \varepsilon_r = \mu_r = \left(\frac{b-a}{b}\right) \tag{4.12}$$

Equations (4.6) and (4.12) show that the entire tensor component has gradients as a function of radius, leading to complex metamaterial design.

The transformation media that forms the cloak lies between two spheres. Figure 4.5 shows a schematic of PEC sphere of radius a = 3 cm and a surrounding cloaking shell of radius b = 11 cm with material properties obtained from optical transformation theory as given in Eq. (4.12)

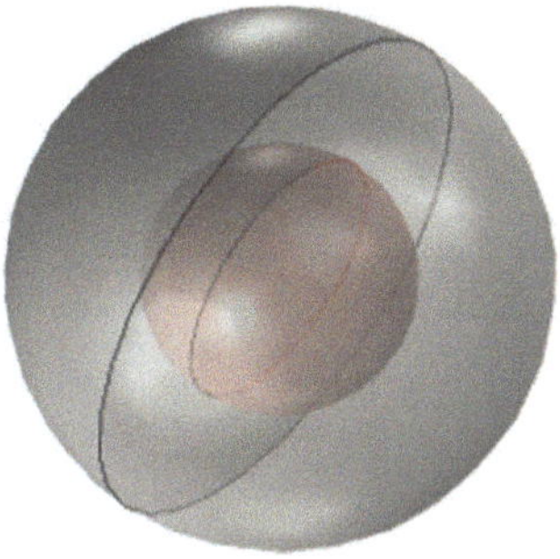

Fig. 4.5 Typical spherical geometry used for simulation of ideal spherical cloak

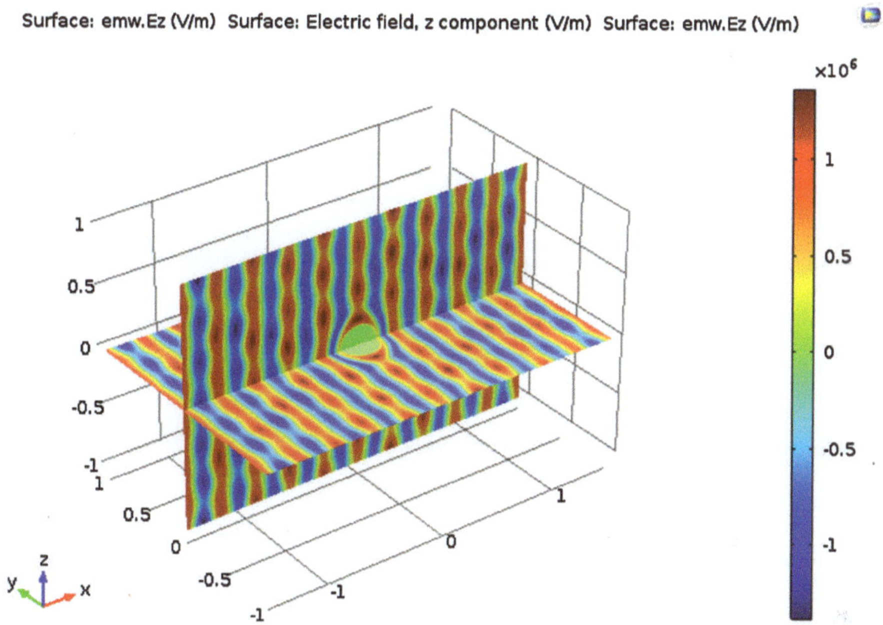

Fig. 4.6 3D visualization of spherical cloak, electric field patterns

The RF and wave optics module has been used to simulate and validate the performance of the spherical cloak. This approach describes the propagation of electromagnetic waves based on numerical integration of set of Hamilton's equations obtained by taking the geometric limit of Maxwell's equations in anisotropic, inhomogeneous media. After deriving the mathematical formulation for the permittivity and permeability tensors, the simulation using COMSOL Multiphysics software has been carried out. The same boundary conditions as mentioned in the previous section have been used for the analysis. Figure 4.6 shows the electric field distribution through the spherical cloak.

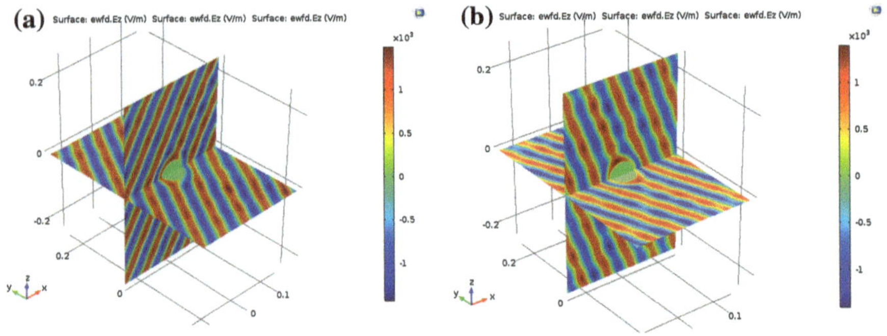

Fig. 4.7 3D visualization of spherical cloak with electromagnetic wave incident at **a** 60°, **b** 250°

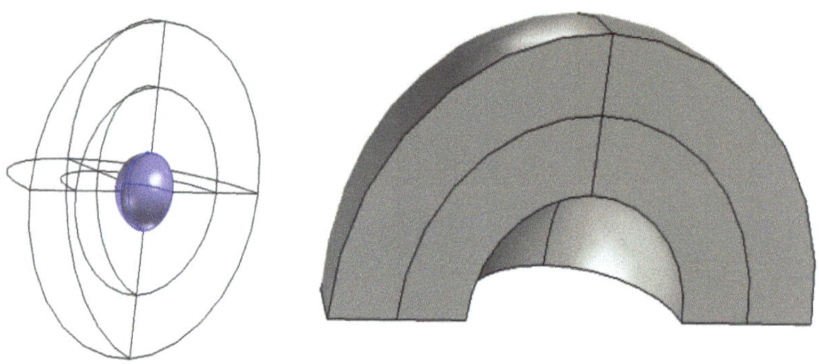

Fig. 4.8 Computational domain for computing the RCS of a PEC sphere in free space

Cloaking effect is also analyzed by projecting the steady-state electric field at different angles. Figure 4.7 shows the simulation of cloaking effect at different angle of incidence using wave optics module of COMSOL Multiphysics software with same boundary conditions as applied to the circular cylindrical cloak.

4.4 RCS Calculation of Multilayer Cloak Structure

Radar cross section is one of the quantitative measures for calculating the performance of an idealized cloak in free space illuminated by a plane wave. In this analysis, the geometrical model used for RCS calculation is a metallic sphere with a very high conductivity, which can be treated as a material with a PEC sphere with radius of 3 cm as shown in Fig. 4.8. Due to symmetry, it suffices to model one quarter of the sphere.

The physics applied to calculate RCS is as follows, scattering object (PEC) is placed at the center of the two concentric spherical shells. Innermost shell adjacent to the PEC is the air domain and is considered as a far-field domain. The outer most shell is the perfectly matched layer (PML) domain which approximate a boundary to infinity. Additionally, scattering boundary condition is applied to the outer most boundary of the structure. For good accuracy, there should be at least five elements for PML domain along the direction to infinity. This condition is efficiently met by using a swept mesh. It consists of a free tetrahedral mesh around the sphere and a swept mesh in the PML domain. When solving scattering problems, the solution converges slowly with respect to the mesh size in the far-field domain i.e., in the region around the scattering object.

The boundary conditions in the model are such that the sphere has perfect electric conductor (PEC) boundaries, thus assuming infinite material conductivity. PEC boundary conditions and perfect magnetic conductor (PMC) boundary conditions apply on the symmetry planes which is used to subdivide the sphere model. The background plane wave travels in the positive x direction, and electric field polarized along z-axis. RCS of PEC sphere is calculated by applying the above-mentioned boundary conditions and the obtained RCS plot is shown in Fig. 4.10. Here a is the radius of PEC sphere and λ is the wavelength corresponding to operating frequency of 3 GHz. Figure 4.9 shows the measured RCS plot of a sphere for reference.

RCS plot of the PEC sphere, shown in Fig. 4.10, has been validated with the measured RCS plot of sphere shown in Fig. 4.9. It can be seen that both the plots are almost similar.

Geometry of one-layer cloaking shell enclosing the PEC sphere is shown in Fig. 4.11. The first layer cloaking shell has 5 cm radius and 0.7 cm thickness. Material properties of the cloaking shell are applied based on the permittivity and permeability tensor components obtained from the transformation theory. The air gap is provided between PEC sphere and cloaking shell. Two concentric circular shells surrounding the cloak are air domain and PML domain, where air domain is used as the far-field domain. RCS analysis of one-layer cloaking structure has been carried out by applying the above-mentioned boundary condition. The obtained RCS of one-layer cloaking shell is as shown in Fig. 4.12, where a is radius of the one-layer cloaking shell and λ is the wavelength corresponding to the operating frequency of 3 GHz.

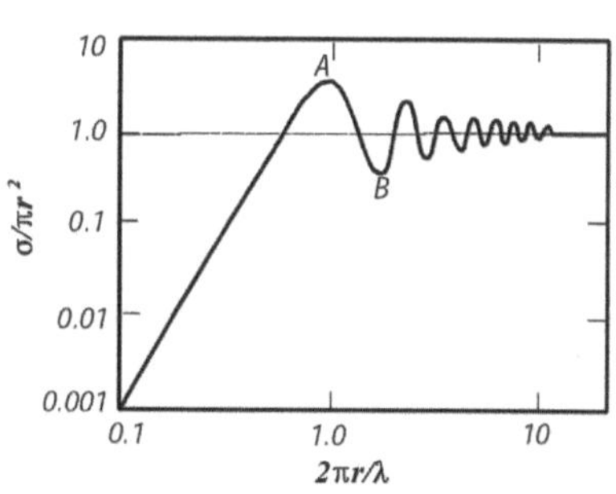

Fig. 4.9 Measured RCS plot of the sphere (c.f. Dr. Allen E. [6])

Fig. 4.10 Simulated RCS plot of the PEC sphere

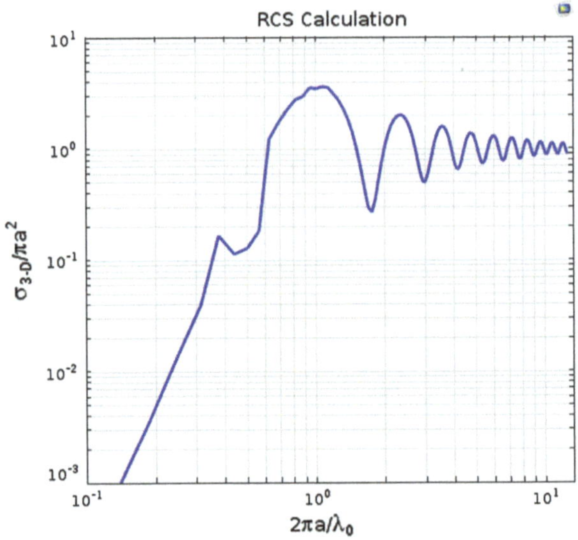

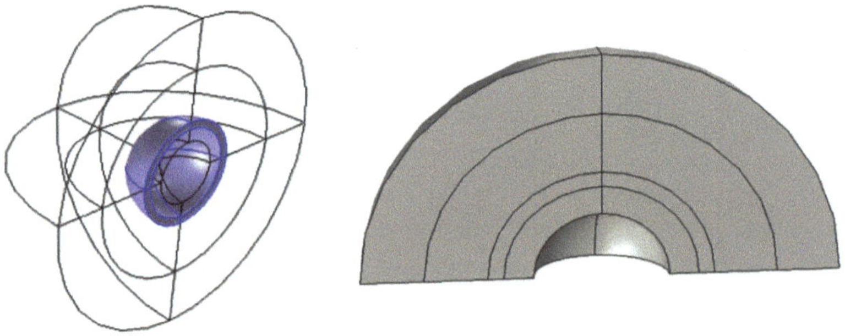

Fig. 4.11 Computational domain for computing the RCS of a one-layer cloaking shell in free space

Fig. 4.12 RCS plot of the one-layer cloaking shell with respect to a/λ (color online)

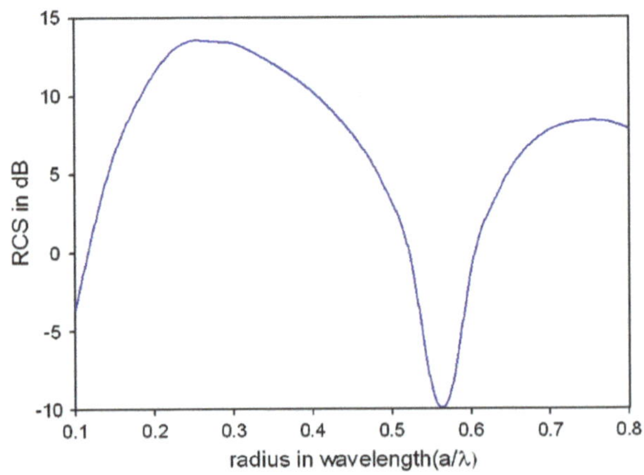

Geometry of two-layer cloaking shell enclosing the PEC sphere is shown in Fig. 4.13. The one-layer cloaking shell has the radius of 5 cm, and outer layer cloaking shell has 7 cm radius, both have the thickness of 0.7 cm. The material properties of the cloaking shell are applied based on the permittivity ε and permeability μ tensor components obtained from the transformation theory. RCS analysis of the two-layer cloaking spherical shell has been carried out, and RCS is calculated by applying the above-mentioned boundary condition. The plot has been reported in Fig. 4.14 where a is the radius of the outer layer cloaking shell and λ is the wavelength with the operating frequency of 3 GHz.

Geometry of three-layer cloaking shell enclosing the PEC sphere is shown in Fig. 4.15. The radii of different cloaking layers such as 1st, 2nd, and 3rd layers are 5, 7, and 9 cm, respectively, with each layer has the thickness of 0.7 cm and air gap is provided between the layers. Material properties of the cloaking shell are applied based on the permittivity ε and permeability μ tensor components obtained from the transformation theory. RCS of the three-layer cloaking shell is calculated by

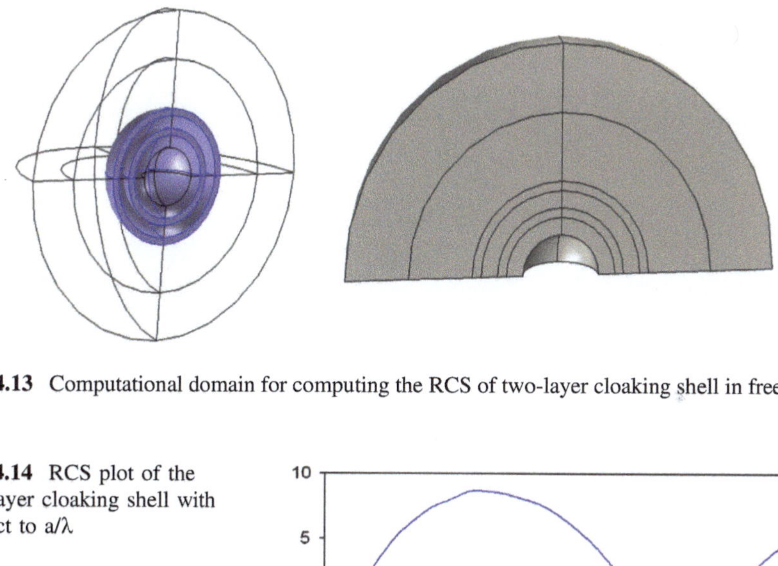

Fig. 4.13 Computational domain for computing the RCS of two-layer cloaking shell in free space

Fig. 4.14 RCS plot of the two-layer cloaking shell with respect to a/λ

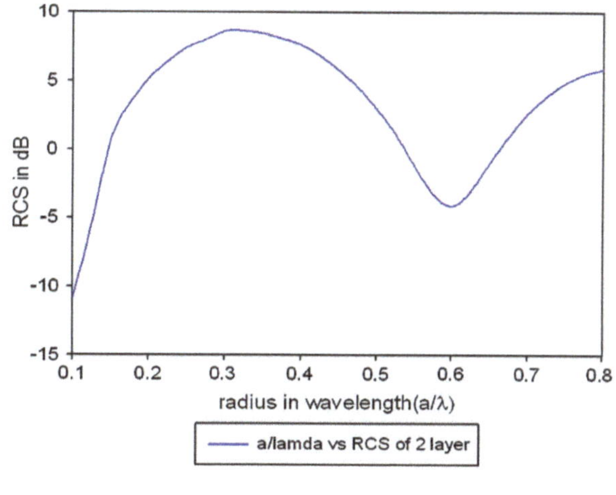

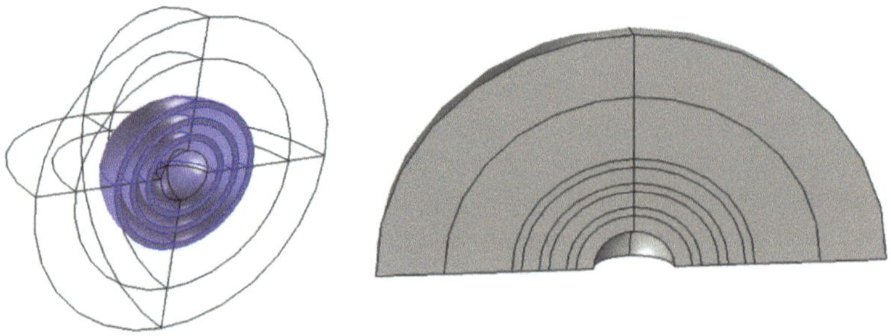

Fig. 4.15 Computational domain for computing the RCS of three-layer cloaking shell in free space

Fig. 4.16 RCS plot of the three-layer cloaking shell with respect to a/λ

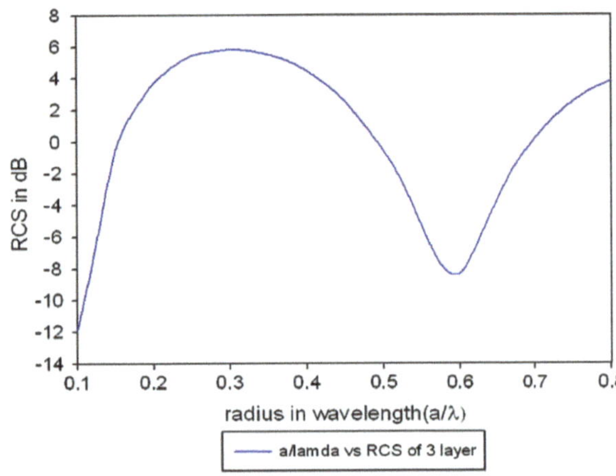

applying the above-mentioned boundary condition, and the plot is as shown in Fig. 4.16 where a is the radius of the 3rd layer cloaking shell and λ is the wavelength corresponding to the operating frequency of 3 GHz.

Geometry of four-layer cloaking shell enclosing the PEC sphere is shown in Fig. 4.17. The radii of different cloaking layers such as 1st, 2nd, 3rd, and 4th layers are 5, 7, 9, and 11 cm, respectively, with each layer having thickness of 0.7 cm and air gap is provided between the layers. Material properties of the cloaking shell are applied based on the permittivity ε and permeability μ tensor components obtained from the transformation theory. RCS analysis of the four-layer spherical cloaking shells has been done, and RCS is calculated by applying the above-mentioned boundary condition. The plot is shown in Fig. 4.18 where a is the radius of the 4th layer cloaking shell and λ is the wavelength corresponding to the operating frequency of 3 GHz.

The RCS analysis of a designed ideal sphere without cloaking shell and with multilayer cloaking shell has been done. The comparison of RCS for sphere in Mie region is represented in Fig. 4.19, it is seen that in the C band there is more

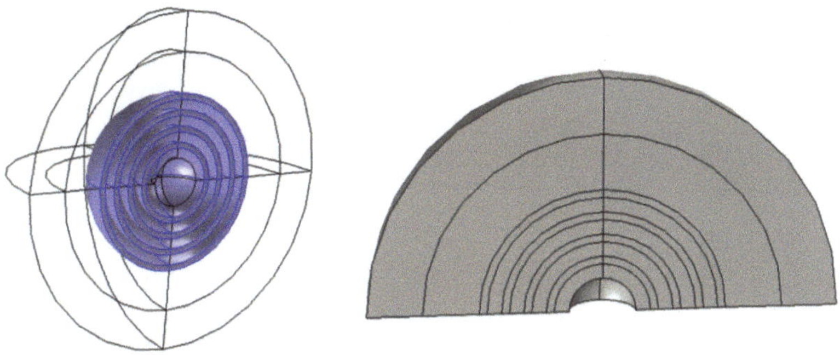

Fig. 4.17 Computational domain for computing the RCS of four-layer cloaking shell in free space

Fig. 4.18 RCS plot of the four-layer cloaking shell with respect to a/λ

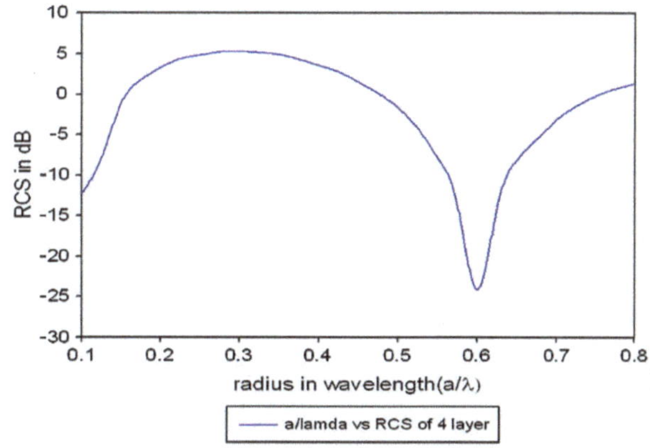

Fig. 4.19 RCS comparison plot between PEC and multilayer cloaking shell with respect to a/λ (color online text)

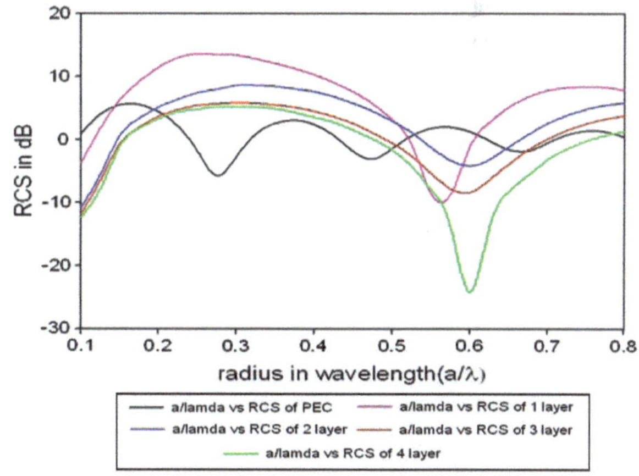

reduction in RCS value hence the operating frequency can be chosen in this band to get the better performance. From Fig. 4.19, it can be seen that, as the number of layers increases there is a reduction of RCS and for the four-layer cloaking shell gives less RCS value as compared to other three cloaking layers.

4.5 Conclusion

In this chapter, design of ideal circular cylindrical cloak and ideal spherical cloak has been demostrated. The detailed mathematical formulation for the electromagnetic material tensors has been reported along with the validation through full wave simulation. The bending of electromagnetic waves has been shown through electric field distribution simulated using RF and wave optics module of COMSOL Multiphysics software. The simulation studies for various angles of incidence have also been reported, and from the analysis it was observed that the electromagnetic field bends around the object and emerges out without any absorption and scattering. Further, towards implementation of the cloaking concept to aerospace domain, a physical insight for calculating the RCS of the multilayer spherical cloak has been provided. The RCS plot of the sphere without cloak and with multilayer cloak has been analyzed, for C band and the reduction in RCS have been observed as the number of layer increased to four. The next chapter of this book gives the design procedure of achieving the formulated permittivity and permeability tensors through metamaterials.

References

1. L. Yu, J. Zhang, H. Chen and B. Wu, "Full-wave analysis of prolate spheroidal and hyperboloidal cloaks," Journal of Physics, vol. 41, pp. 1–6, 2008.
2. D. Schurig, J. J. Mock, B. J. Justice, S. A. Cummer, J. B. Pendry, A. F. Starr, and D. R. Smith, "Metamaterial electromagnetic cloak at microwave frequencies," Science, vol. 314, pp. 977–980, 2006.
3. B. Pendry, D. Schurig, and D. R. Smith, "Controlling electromagnetic fields," Science, vol. 312, pp. 1780–1782, 2006.
4. Q. Cheng, W. X. Jiang, and T. J. Cui, "Investigations of the electromagnetic properties of three-dimensional arbitrarily-shaped cloaks," Progress in Electromagnetics Research, vol. 94, pp. 105–117, 2009.
5. B. Choudhury, P. V. Reddy, R. M. Jha, Permittivity and Permeability Tensors for Cloaking Applications, Springer, New York, ISBN:978-981-287-805-2, p. 71, 2016.
6. A. E. Fuhs, Radar Cross Section Lectures, Department of Aeronautics, Naval Postgraduate School, California, p. 159, 1982.

Chapter 5
Design Optimization of Cloaks

Pavani V. Reddy, Susan Thomas and Balamati Choudhury

5.1 Introduction

The development in metamaterial science and technology has created many exhilarating applications in microwave to optical frequency region in which invisibility cloaking [1] is one of the exciting applications. In recent years, the RCS reduction characteristics of cloaking structures have been widely investigated as it found extensive applications in stealth platform. Maxwell's equation with transformation theory [2] is the fundamental principle behind the operation of the invisibility cloak in anisotropic medium. The design of invisibility cloak for arbitrary shape is the optimal version of EM stealth property in which the electromagnetic waves are controlled and guided within the cloaking shell by applying the prescribed spatial variation in the constitutive parameters. The spatial variation in the material properties like relative permittivity and permeability is performed using the coordinate transformation, resulting in bending of EM waves around the object. Calculation of the relative permittivity and permeability tensors [3] is one of the promising challenges in the design of invisibility cloak. In aerospace applications, invisibility implies hiding information about the target from detecting objects like radar. This can be achieved by reducing the radar cross section (RCS) of the object by using various stealth techniques and as a result, achieving no absorption and reflection of the energy by the object [4–6].

P.V. Reddy (✉) · S. Thomas · B. Choudhury
Centre for Electromagnetics, CSIR-National Aerospace Laboratories,
Old Airport Road, Bengaluru 560017, India
e-mail: Pavani.paluri541@gmail.com

B. Choudhury (ed.), *Metamaterial Inspired Electromagnetic Applications*,
DOI 10.1007/978-981-10-3836-5_5

5.1.1 Application of Tensors in Electromagnetic Cloaking

Tensors find an important niche in diverse areas of applications such as continuum mechanics, aerospace structural analysis, quantum computing and quantum mechanics. Recently, researchers have been attracted to an interesting electromagnetic phenomenon which may pave the way for invisibility in near future. The basic concepts and understanding of tensors and their analysis is an inevitable area of research in electromagnetic invisibility cloaking and details about tensors has been given in the Chap. 3 of this book. Any material is being characterized by its unique properties such as electric permittivity and magnetic permeability. Another property called refractive index can be obtained from permittivity and permeability. Since dielectric permittivity and permeability of the cloaking layers are spatially varying and anisotropic, they are called as tensors. For example, if a cylindrical cloak is to be designed, the dielectric permittivity and permeability have three components along r, θ, z which varies as a function of radial distance.

5.2 Significance of Metamaterials in Cloaking

Cloaking is an electromagnetic phenomenon which can bend the EM waves in an unusual way. This phenomenon cannot be achieved with the help of natural dielectrics. Thus, artificial periodic structures are being implemented for achieving electromagnetic cloaking at microwave frequencies. Air has dielectric permittivity and permeability equal to unity. Dielectric permittivity for naturally existing materials is found to be greater than unity. Most of the materials have magnetic permeability equal to unity. The dielectric permittivity and permeability for cloaking applications are found to be between zero and unity or more specifically, near to zero. Metamaterials are artificial repetitive structures that can exhibit electromagnetic properties which are not present in natural materials.

An important peculiarity of metamaterials is that the material properties can be varied by changing the structural properties as per the requirement or specifications for various applications. Therefore, metamaterials which exhibit unusual properties were adopted in the design of cloaking shells or layers. The object can be concealed by flexible metamaterial layers with varying constitutive material properties. These spatially varying anisotropic material properties can be termed as tensors. The electromagnetic wave paths can be controlled by the use of appropriate tensors and therefore, the study of electromagnetic tensors becomes significant in metamaterial-based cloaking. In short, metamaterials provide unprecedented control over the electromagnetic properties.

5.3 Design of Metamaterial Unit Cells for Cloaking Applications

The term "invisibility cloaking" at microwave frequencies is attributed to reduction in the forward and backward scattering from the object which is concealed. The first experimental realization of electromagnetic cloak is achieved by Schurig. The properties of metamaterials have been utilized in a proper way to achieve invisibility at microwave frequencies. The reason for selection of metamaterials for cloaking applications has been already discussed in the last section.

According to the coordinate transformation principle in cloaking, though the Maxwell's equations remain invariant to the transformation, the material properties such as permittivity and permeability tensors undergo a variation, which is a function of the spatial coordinates. Assume that a cylindrical object is to be hidden. Applying coordinate transformation for a cylindrical object to be hidden, the cylindrical region $0 < r < b$ needs to be compressed to an annular region $a < r' < b$. The expressions for the tensors on applying coordinate transformation for a cylindrical cloak in transformed media as well as reduced media are given in Table 5.1. Only three tensors are relevant because of the field excitation considerations for the cloak design which will be discussed in the next section. It can be observed that the transformed material properties are complex and depend on the radius of the cloaking layer, inner and outer radius of the cloak. These equations reveal a complicated cloak design. By compromising for a zero reflected power, the properties can be simplified. The reduced material properties provide more flexibility in the design of cloak. In reduced material specifications, two tensors remain constant, whereas the third varies depending upon the radius of the cloaking layer. Also, the reduced media holds the same dispersion relation as well as wave paths in the cloak as that of fundamental transformed properties.

Though permeability and permittivity tensors in certain spatial coordinate remain constant, permeability tensor in another spatial coordinate varies depending upon the distance from the object to be hidden. Therefore, the metamaterial structure chosen for cloaking application should have a property so as to tailor the magnetic response depending on the geometrical parameters of the unit cell. Split-ring resonator (SRR) is a widely adopted metamaterial unit cell structure for various applications

Table 5.1 Expression for permittivity and permeability tensors on application of coordinate transformation

Tensors (permittivity and permeability)	Transformed material properties	Reduced material properties
μ_r	$\frac{r-a}{r}$	$\left(\frac{r-a}{r}\right)^2$
μ_θ	$\frac{r}{r-a}$	1
ε_z	$\left(\frac{b}{b-a}\right)^2 \frac{r-a}{r}$	$\left(\frac{b}{b-a}\right)^2$

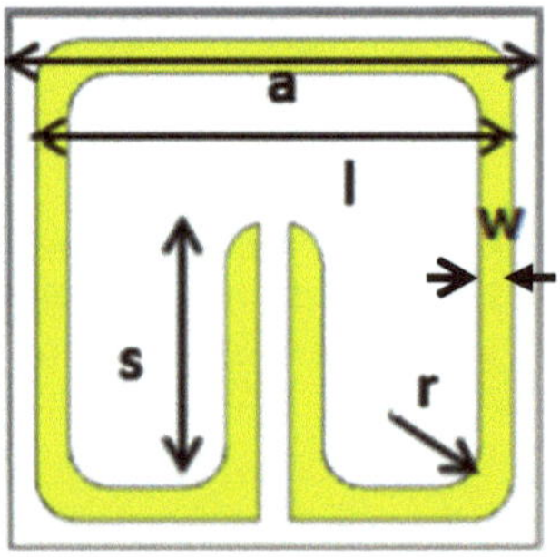

Fig. 5.1 Design of SRR unit cell

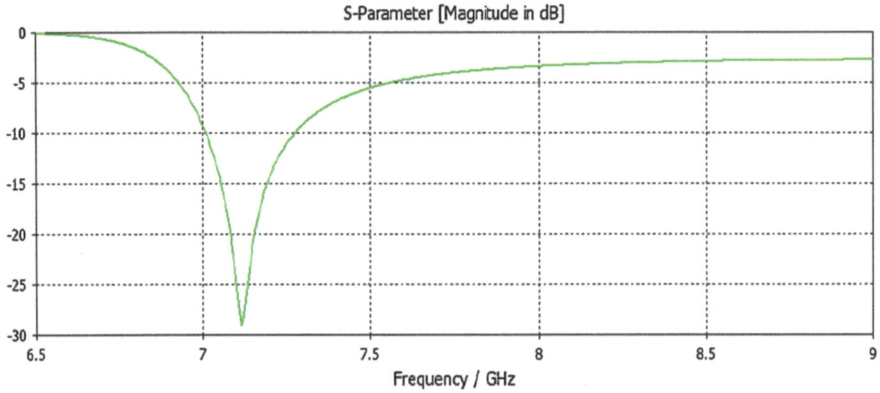

Fig. 5.2 Magnitude of transmission parameters (S_{21}) for SRR unit cell

such as antennas, absorbers and lenses. SRR exhibits a magnetic resonance that can be altered by varying the geometrical parameters of the unit cell. SRRs can be implemented in square as well as circular shapes. For electromagnetic cloaking applications, a slight modification has been made in the geometrical structure of square-shaped SRR. The split in the square-shaped ring is elongated toward inside of the square-shaped ring and the corners of the SRR are kept curved.

Figure 5.1 shows the schematic of SRR unit cell on Duroid 5870 substrate of dielectric constant of 2.33 and loss tangent of 0.0012 at 10 GHz. The thickness of the Duroid substrate is 0.381 mm, and the split-ring resonators are patterned with a thickness of 0.017 mm.

Widely accepted FEM-based EM simulation tools have been used for the design and simulation of metamaterial unit cell. The frequency range selected for simulation is 6.5–9 GHz. Figures 5.2 and 5.3 show the magnitude and phase of transmission

Table 5.2 Dimensions of SRR unit cell (in millimeters)

a	10/3 mm
l	3 mm
w	0.2 mm
r	0.26 mm
s	1.654 mm

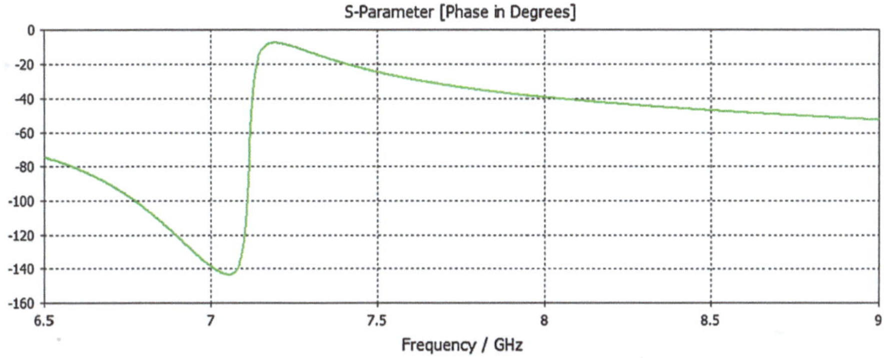

Fig. 5.3 Phase of transmission parameters (S_{21}) for SRR unit cell

characteristics of the SRR unit cell which resonates at 7.116 GHz respectively. The first layer of cylindrical cloak has been designed using unit cells of dimensions mentioned in Table 5.2.

The length of the split and radius of the corner are the two design parameters which affect the magnetic response of the SRR.

The realization of an entire metamaterial structure requires the design of periodic array of unit cells. Such structures can be simulated by applying appropriate boundary conditions to a single three-dimensional unit cell. This helps in reducing the tedious task of repeated design and simulation of unit cells. The unit cell is excited with a plane wave with electric field in a direction parallel to the SRR plane and magnetic field normal to the surface of unit cell. Since the electric field is parallel to unit cell plane, the capacitive component due to split in the SRR unit cell is not excited directly.

5.4 Scattering Parameter-Based Retrieval of Material Properties of Metamaterial Unit Cell

Material parameters of metamaterial unit cells such as permittivity and permeability can be extracted from scattering parameter simulation results obtained from EM simulation tool. MATLAB script has been used to compute the complex permittivity and permeability from the scattering parameters. The S parameter-based retrieval is a reliable technique for material parameter extraction. This help in solving the inverse problems to design the metamaterial unit cells according to the specifications of permittivity and permeability for various applications.

The plots obtained for variation of real and imaginary parts of material properties such as electric permittivity and magnetic permeability depending on the frequency are depicted in Figs. 5.4, 5.5, 5.6, and 5.7. The real part of permittivity is always positive in the frequency range selected for simulation, i.e., 6.5–9 GHz. The imaginary part of permittivity and permeability is near to zero except at resonance. The real part of permeability shows an abrupt change in the value around the resonance frequency. The permeability reaches near to zero from a frequency of 7.5–9 GHz.

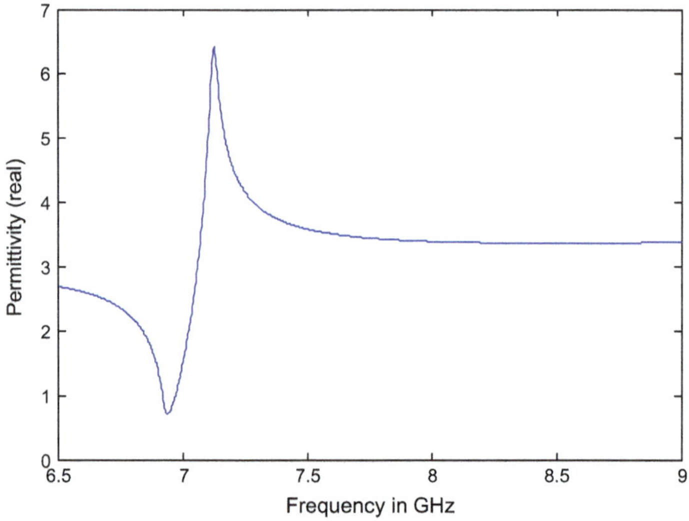

Fig. 5.4 Extracted permittivity (real) versus frequency

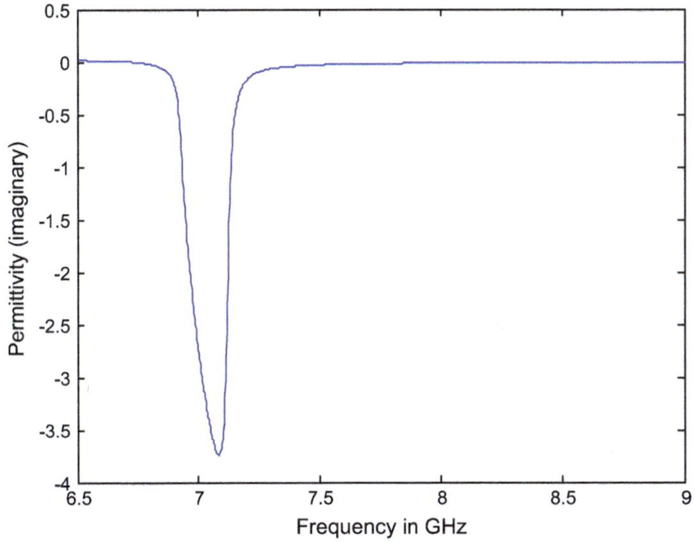

Fig. 5.5 Extracted permittivity (imaginary) versus frequency

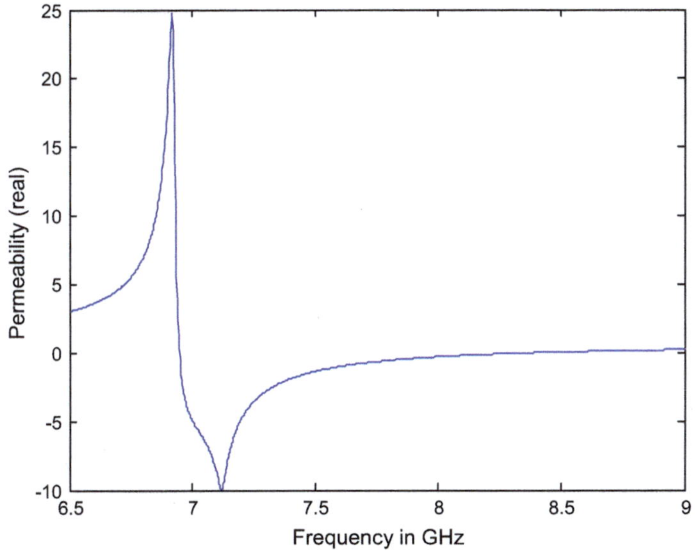

Fig. 5.6 Extracted permeability (real) versus frequency

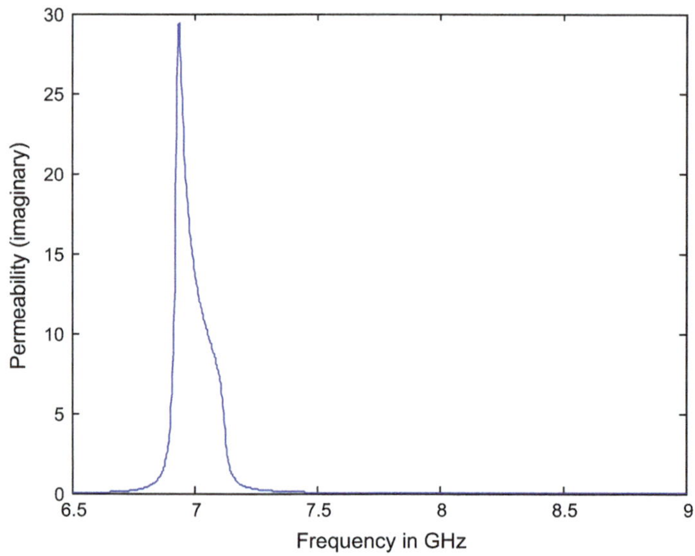

Fig. 5.7 Extracted permeability (imaginary) versus frequency

5.5 Resonance Splitting in Metamaterial Cloak

For an infinite two-dimensional array of resonators, periodic boundary conditions are applied along the directions normal to wave propagation. In this case, the resonators are excited in phase. Since the simulation of the entire cloaking structure is practically impossible, small sections of cloak are analyzed to get an idea about the nature of the electromagnetic cloak. Two types of metamaterial unit cell arrays can be designed: linear and planar.

Linear array of metamaterial unit cells

The SRR unit cells can be arranged in a linear fashion in 3 different ways. An analysis of the transmission parameters for each case has been discussed in the succeeding section.

Case 1: Metamaterial unit cells arranged along the direction of electric field

The stacking of SRR unit cells along the direction of electric field is shown in Fig. 5.8. This results in a strong resonance splitting with a bandwidth of nearly 1 GHz as shown in Fig. 5.9. These simulations are performed using plane wave incidence in free space.

Case 2: Metamaterial unit cells arranged along the direction of propagation of EM wave

The excitation of resonator array stacked along the direction of wave propagation vector results in a wave delay and phase difference between the elemental excitations. This phase difference results in splitting of the resonance. When the SRRs are

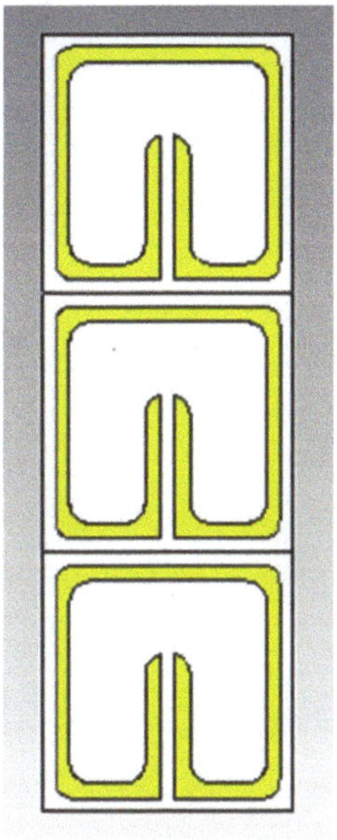

Fig. 5.8 Metamaterial unit cells stacked along the direction of electric field

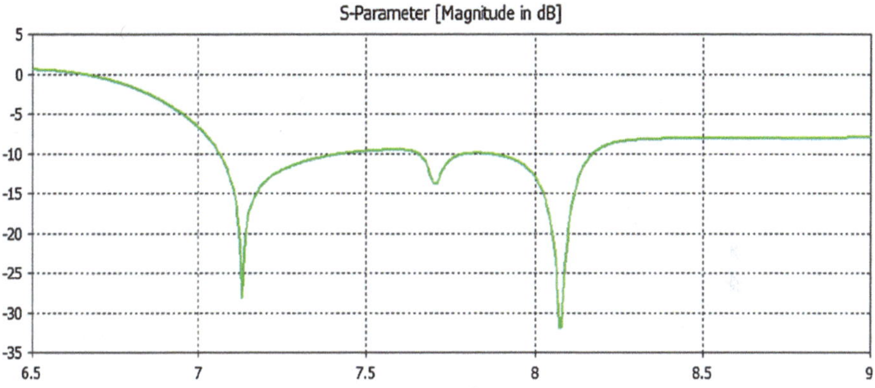

Fig. 5.9 Transmission (S_{21}) parameters of metamaterial array

arranged periodically along the direction of propagation of EM wave, a split is observed in the resonant frequency. More the number of unit cells stacked along the direction of propagation wave vector, the larger will be the separation between the multiple frequency dips. The arrangement of cells is shown in Fig. 5.10 and the simulation results are presented in Figs. 5.11 and 5.12.

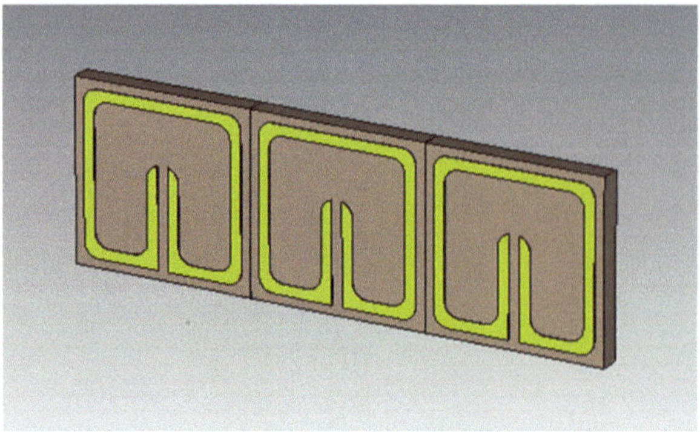

Fig. 5.10 Metamaterial unit cells stacked along the direction of wave propagation

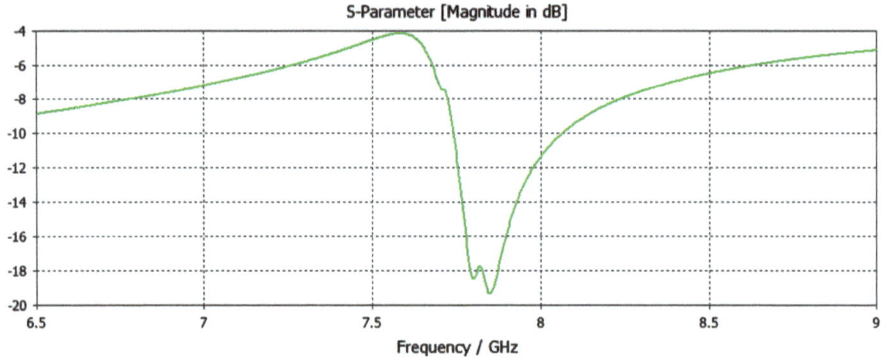

Fig. 5.11 Transmission (S_{21}) parameters of metamaterial array

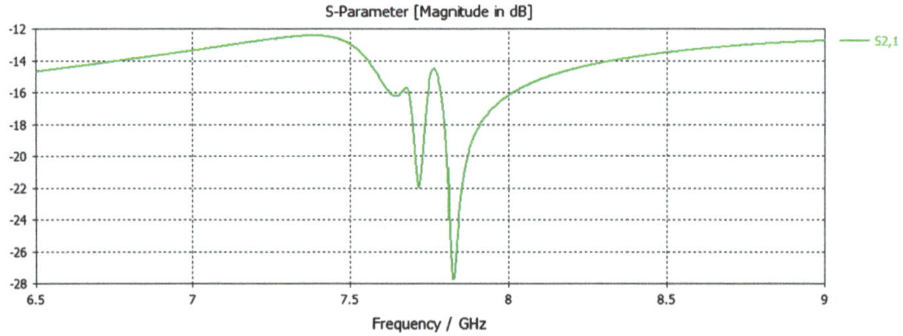

Fig. 5.12 Splitting of resonance in an array of 5 unit cells stacked along the wave propagation vector

Case 3: Metamaterial unit cells arranged along the direction of electric field with orientation reversed alternatively

The two-dimensional cylindrical cloak has a height of three unit cells and the layers are arranged cylindrically. The SRR unit cells are oriented alternatively to reduce the magnetoelectric coupling inherent in the single-split SRR. The orientation of unit cells is shown in Figs. 5.13 and simulated transmission characteristics is shown in Fig. 5.14.

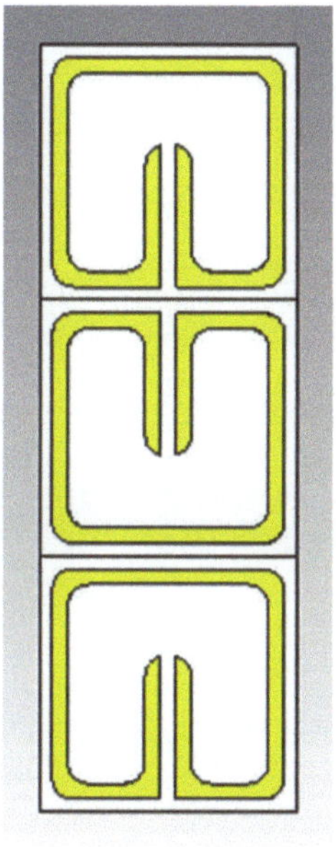

Fig. 5.13 Metamaterial unit cells stacked along the direction of electric field with alternative orientation

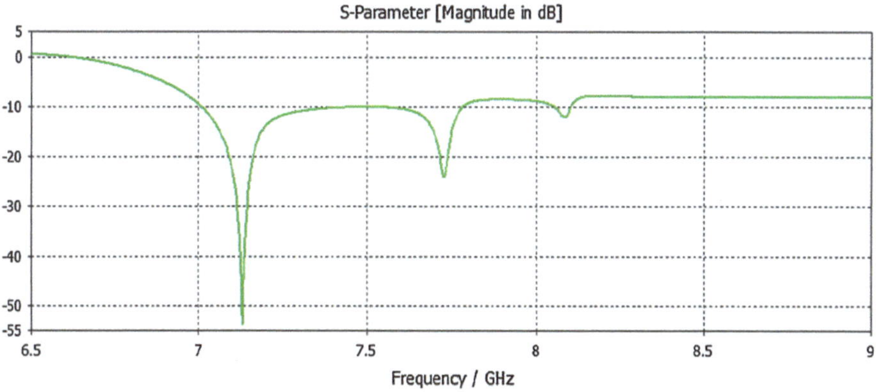

Fig. 5.14 Transmission (S_{21}) parameters of metamaterial array shown in Fig. 5.13

5.6 Finite Planar Array of Metamaterial Unit Cells

A finite array of three unit cells along both directions has been designed to observe the transmission parameters of a finite metamaterial array used for cloaking applications. The schematic of a 3 × 3 array of SRR unit cells is shown in Fig. 5.15. The S_{21} parameters obtained for such an array are shown in Fig. 5.16.

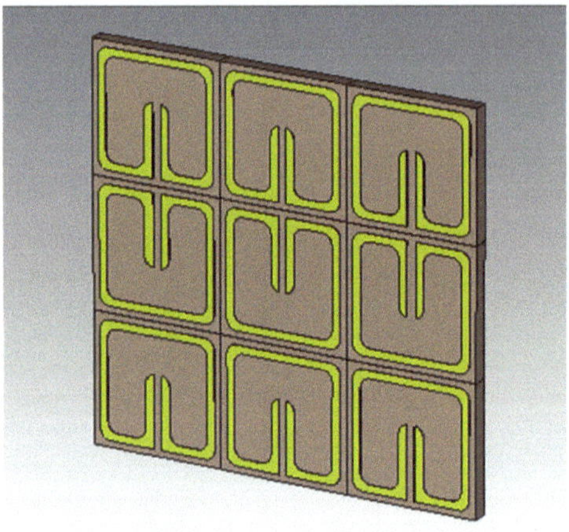

Fig. 5.15 Finite two-dimensional metamaterial array (3 × 3)

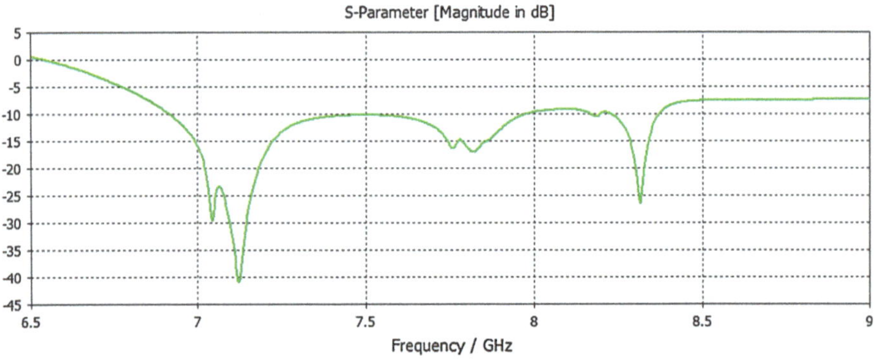

Fig. 5.16 Transmission (S_{21}) parameters of metamaterial array shown in Fig. 5.15

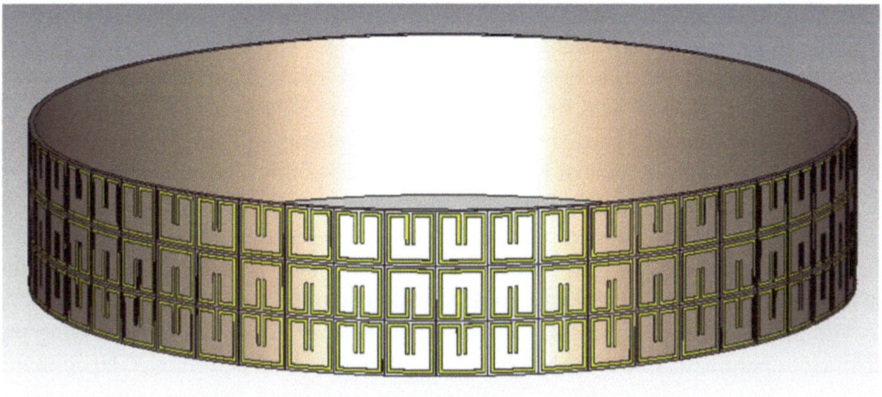

Fig. 5.17 Design of a layer of the cylindrical microwave cloaking structure with SRR unit cells

5.7 Two-Dimensional Cylindrical Cloak

The first ever implemented two-dimensional cylindrical cloak is composed of ten layers of cloaking shells. Each cylindrical layer consists of identical unit cells. The cloak structure has a height of three unit cells. For each cylindrical layer, the geometrical parameters of unit cells are slightly varied to obtain a varying magnetic permeability. Figure 5.17 shows the design of the innermost layer of the cylindrical cloak. As the number of cloaking layers increases, the forward and backward scattering reduces.

5.8 Design of Conformal Metamaterial Unit Cells for Invisibility Cloaking Applications

Electromagnetic wave propagation analysis is possible only with the help of basic tensor parameters like permittivity and permeability which varies according to the direction. Effective medium approach in general relativity is used to derive the permittivity and permeability tensors. Being the simplest as well as the basic structure, the formulation of the relative parameters for right circular cylinder and sphere has been done. Extensive investigations have been carried out to design a perfect invisibility cloak which is capable of bending the EM wave around the object to be hidden.

The major challenges in the cloak design are complexity in the realization of cloaking structure and computation of permittivity and permeability tensors for materials filled in the cloaking shell.

Abundant electromagnetic functionalities have emerged with a tremendous growth in the advent of metamaterial composites. Realization of cloaking at microwave frequencies has been investigated from several years, which demands the fabrication of complex multilayer composites. However, the fabrication of sub-wavelength unit cells becomes increasingly challenging while moving from the microwave to visible region of the electromagnetic spectrum. To date, the majority of aforementioned work has been carried out on planar composites. At terahertz (THz) frequencies and above, construction of multiple unit cell structures in the direction of propagation and realization of non-planar metamaterial composites necessitates the development of new fabrication strategies.

The most challenging part in the analysis and design of the cloak is finding the distinctive parameters of metamaterial unit cell used, which determines the cloaking features. Analysis of conformal metamaterial structures becomes inevitable in design and realization of cloaking structures for aircraft and automobile platforms. Several electromagnetic simulations are performed in order to provide a comparative analysis of arbitrary conformal metamaterial unit cell structures with their corresponding planar structures and verified the advantages and disadvantages of each case. The various conformal unit cell structures designed using EM simulation software have been discussed in this section.

5.8.1 I Shape

The I-shaped unit cells are one of the gradient index metamaterial structures for low loss and broadband applications. Figure 5.18a shows the geometry of planar

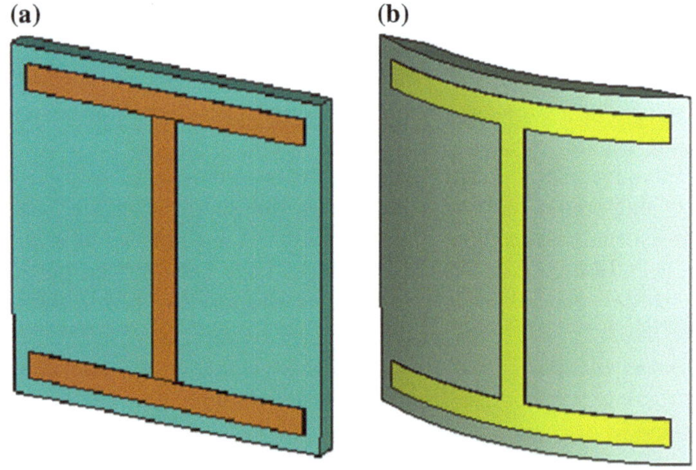

Fig. 5.18 **a** Typical proposed planar I-shaped unit cell. **b** Typical designed conformal I-shaped unit cell

I-shaped unit cell. The I-shaped metallic structure made of gold with, length l = 3.6 mm, line width w = 0.3 mm, and depth t = 0.018 mm is embedded on the polyimide substrate. The dimensions of unit cell are 4 mm × 4 mm × 0.25 mm, respectively. The I-shaped unit cell is conformed onto a cylindrical sector as shown in Fig. 5.18b. The cylinder of radius 2 mm is designed, and a sector of the desired substrate dimension is etched out. Dimensions of metallic patch remains same as that of planar one.

The simulation studies of the conformal unit cell have been carried out using a FEM-based simulation software. The background and boundary conditions of the unit cell were set up properly in order to make the wave propagation and polarization in the desired direction. Also, two waveguide ports are defined and excitations are applied normal to the substrate along the z-direction for obtaining the S parameters. Figures 5.19 and 5.20 represent the reflection and transmission characteristics of the designed planar and conformal unit cell structures, respectively. The scattering parameters obtained for conformal structure show the same trend corresponding to planar one.

The material properties of designed unit cell are extracted from the S parameters using well-developed retrieval method, for analyzing the feasibility of the same for cloaking applications. Figures 5.21 and 5.22 show the retrieved permittivity and permeability of the designed unit cells, respectively. From the figures, it is clear that the designed conformal I-shaped unit cell exhibits negative permittivity and permeability values over X band as similar to planar one.

Cloaking application demands for structures exhibiting negative refractive index, so that the ray propagates in the desired direction. The refractive index of the

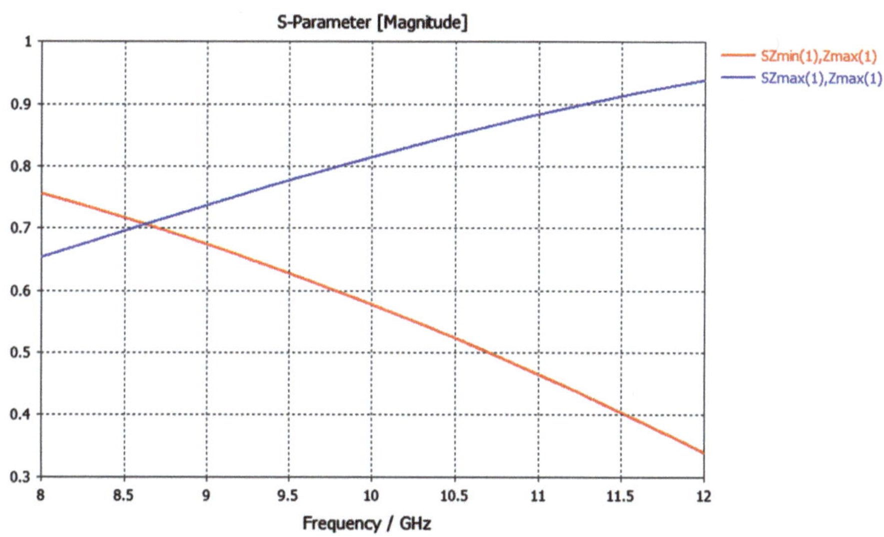

Fig. 5.19 Scattering parameters (S_{11} and S_{21}) of planar unit cell structure of proposed planar I shape

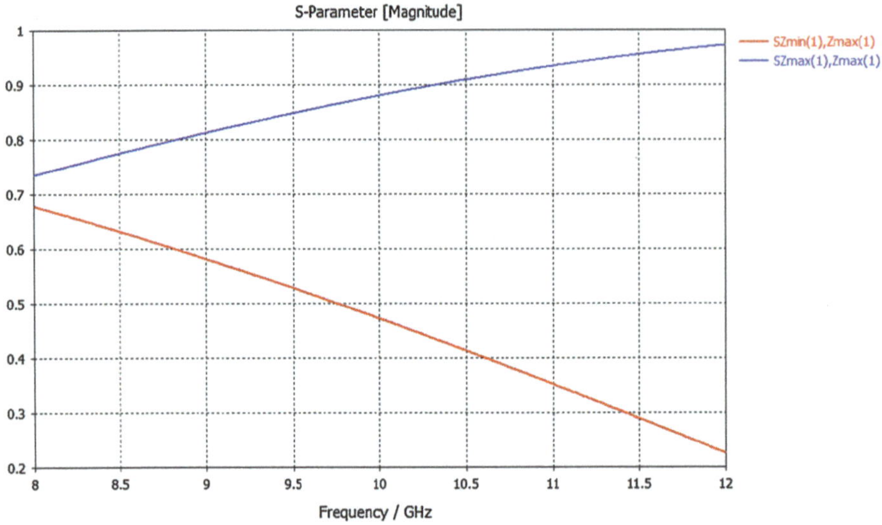

Fig. 5.20 Scattering parameters (S_{11} and S_{21}) of designed conformal I-shaped unit cell structure

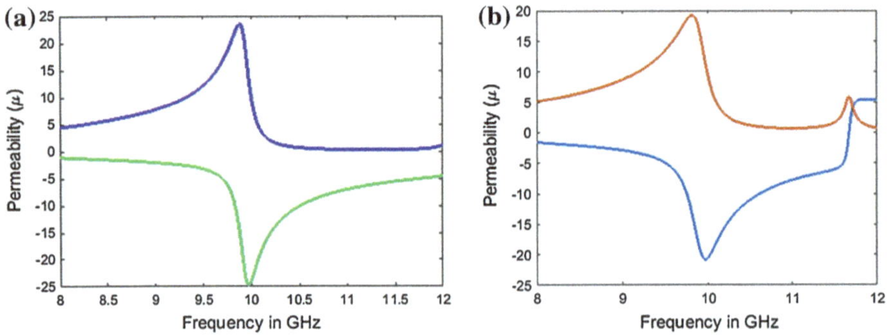

Fig. 5.21 **a** Permeability characteristics of planar I-shaped unit cell. **b** Permeability characteristics of conformal I-shaped unit cell

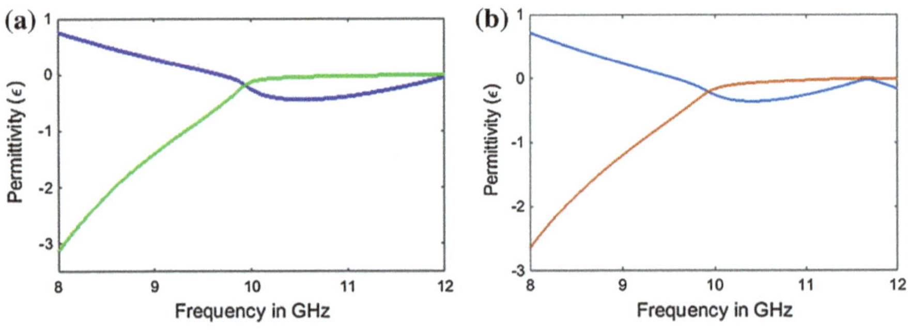

Fig. 5.22 **a** Permittivity characteristics of planar I-shaped unit cell. **b** Permittivity characteristics of conformal I-shaped unit cell

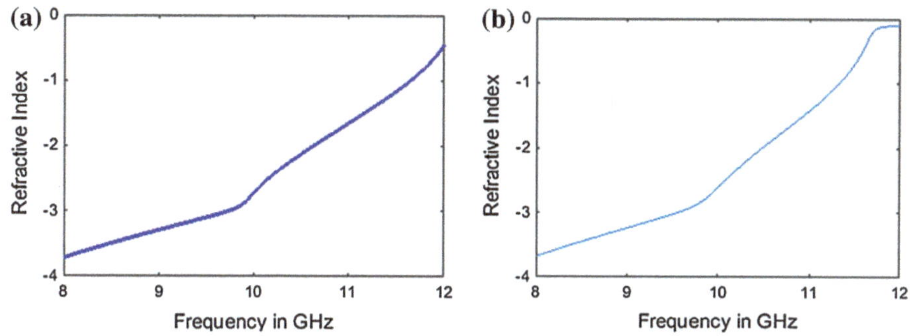

Fig. 5.23 Refractive index of proposed **a** planar I-shaped unit cell, **b** conformal I-shaped unit cell

designed unit cell is calculated from the obtained permittivity and permeability values and is shown in Fig. 5.23. The refractive index values are negative as like planar structure. The designed conformal I structure is a good candidate for cloaking applications, especially for making the targets with curved surfaces to be invisible.

The low loss, broadband, and I-shaped unit cell structure with gradient index is designed and simulated, which is worthy in military aerospace applications. The design can be easily extended to other microwave or optical devices like directional cloaks, beam shifters, beam steering devices, lenses, etc.

5.8.2 H Shape

A conformal H-shaped unit cell with dual-band response is designed and compared with the planar H-shaped unit cell proposed [5]. The proposed H-shaped unit cell is designed by two oppositely wound C-shaped arms aside and a capacitive gap in the center. The C-shaped arms act as inductor and the capacitive gap acts as a capacitor. The unit cell dimension is given by 158 × 168 μm and is designed for terahertz applications. The H-shaped unit cell is conformed into the cylindrical sector as shown in Fig. 5.24. The cylinder of radius 2 mm is designed and a sector of the desired substrate dimension is etched out. Dimensions of metallic patch are same as that of planar one.

The designed structure comprises of two layers, bottom substrate layer is made of SiO_2 and the metallic H-shaped patches made of gold. In order to achieve EM resonance in THz regime, the metallic sub-wavelength unit cell structure are arranged periodically in two perpendicular directions. In this section, a dual-band terahertz metamaterial based on a hybrid "H-"shaped cell with different sizes is designed and simulated, and each cell is constructed by connecting two oppositely wound C-shaped arms aside and a capacitive in the center. The simulation studies

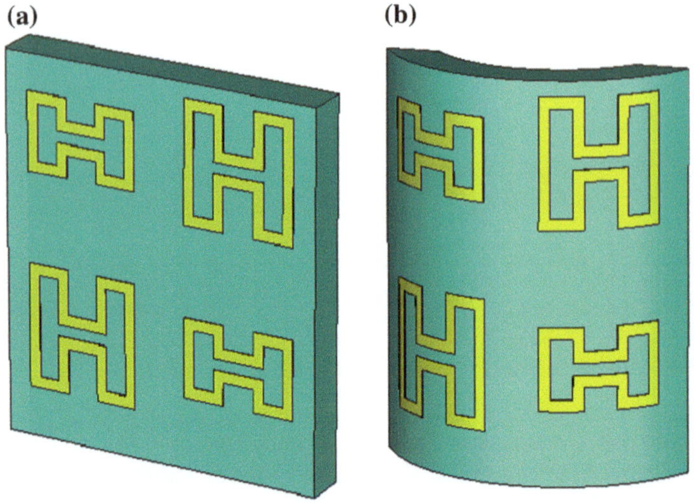

Fig. 5.24 **a** Typical proposed planar H-shaped unit cell. **b** Typical designed conformal H-shaped unit cell

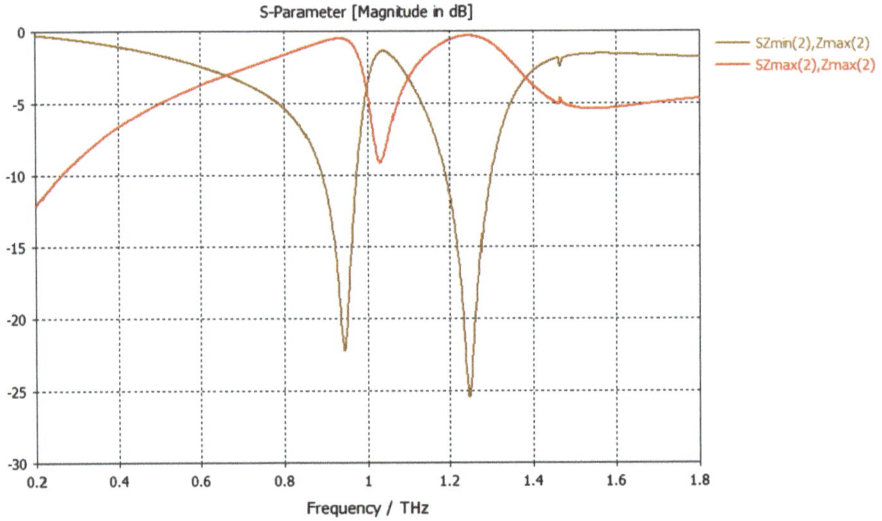

Fig. 5.25 Scattering parameters (S_{11}) and (S_{21}) characteristics of proposed planar H shape

of the conformal unit cell have been carried out using a FEM-based simulation software.

Figures 5.25 and 5.26 show the reflection and transmission characteristics of proposed planar H and designed conformal H-shaped unit cell structure, respectively. Designed conformal H-shaped structure gives better return loss characteristics as compared to planar structure and also it exhibits multiple resonances in the

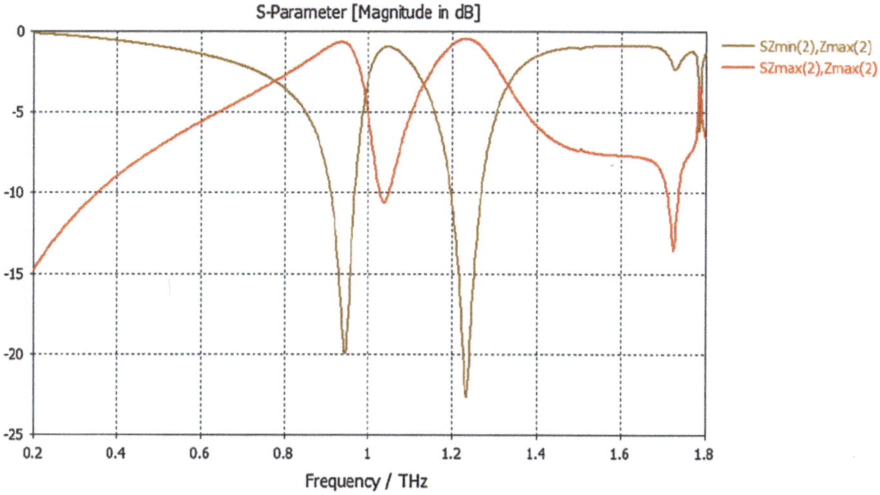

Fig. 5.26 Scattering parameters (S_{11} and S_{21}) of designed conformal H shape

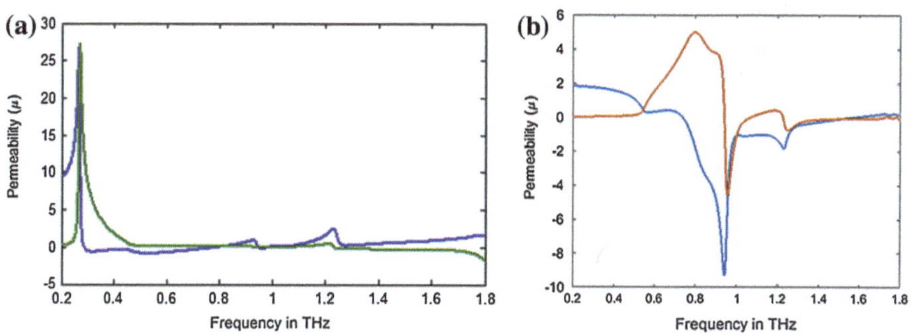

Fig. 5.27 **a** Permeability characteristics of proposed planar H shape. **b** Permeability characteristics of designed conformal H shape

THz regime. Designed structure has less transmission as that of planar one, due to the effect of curvature. From the simulation results, it can be seen that the two electrical resonances at 0.95 THz and 1.26 THz have been occurred as the THz wave propagates normally to the metallic array and THz electric field is perpendicular to the split gap.

The material properties such as permittivity and permeability are calculated from the obtained S parameters and are plotted as shown in Figs. 5.27 and 5.28. The graph summarized that designed structure has better negative characteristics of permittivity and permeability as compared to the planar structure and is worthy to design the invisibility cloak. The designed conformal H-shaped structure which operates in multiple frequencies along with their negative material characteristics makes them, a suitable candidate for terahertz applications.

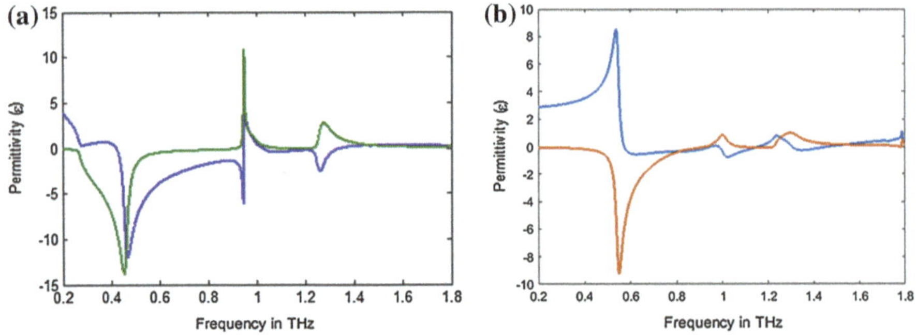

Fig. 5.28 **a** Permittivity characteristics of proposed planar H shape. **b** Permittivity characteristics of designed conformal H shape

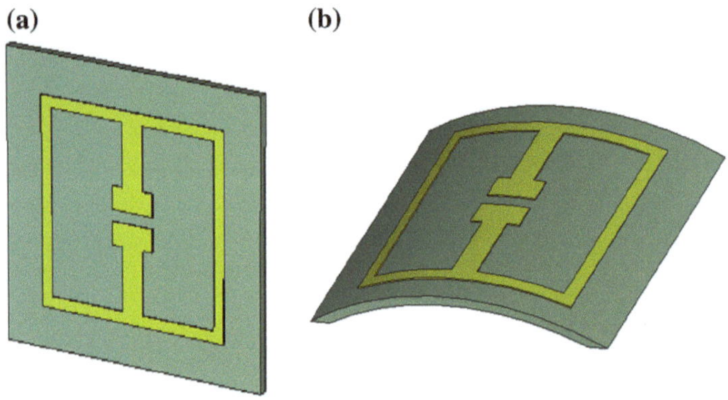

Fig. 5.29 **a** Typical proposed planar RAM-type unit cell. **b** Typical designed conformal H shape

5.8.3 RAM-Type SRR

This section explains the design of conformal metamaterial unit cell structure, which is an extension to the structure proposed [4]. The unit cell designed looks like a RAM absorber, having a center gap which acts as a capacitor in effect. In order to make the structure conformal, the designed metallic patch is etched on a curved substrate. The substrate is made of polyimide with dielectric constant of 4.8. The ultra-thin polyimide acts as low index, low loss, and highly flexible substrate in the fabrication of THz metamaterials. The structure is designed for terahertz frequencies, and the optimum dimensions for the unit cell obtained are 50 μm × 50 μ m × 1.5 μm. The metallic patch has outer dimension of 36 μm, line width of 2 μm. The gap between two middle rods is 0.5 μm, which has much significance as it determines the capacitance value. The designed structure along with planar corresponding is shown in Fig. 5.29.

The designed structure is simulated using a FEM-based EM simulation software, and the obtained reflection and transmission characteristics are shown in

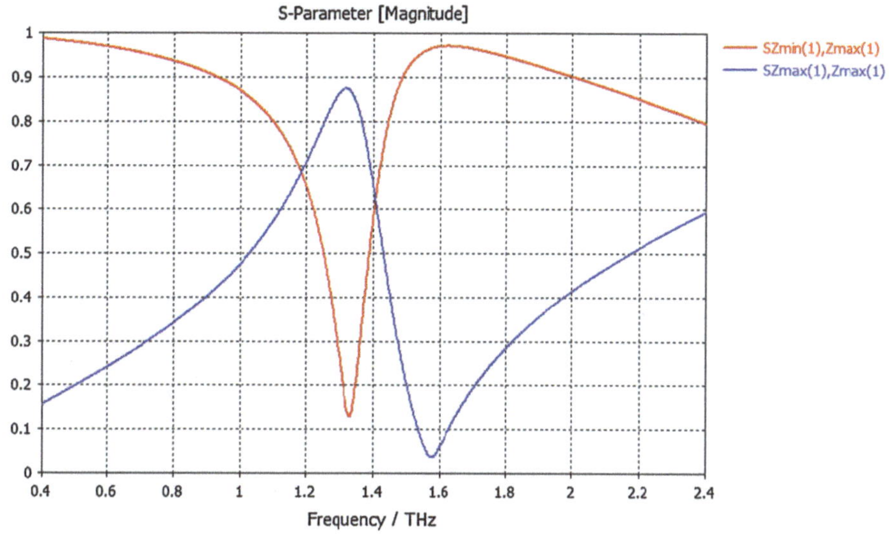

Fig. 5.30 **a** Scattering parameters (S_{11}) and (S_{21}) characteristics of proposed planar RAM-type unit cell

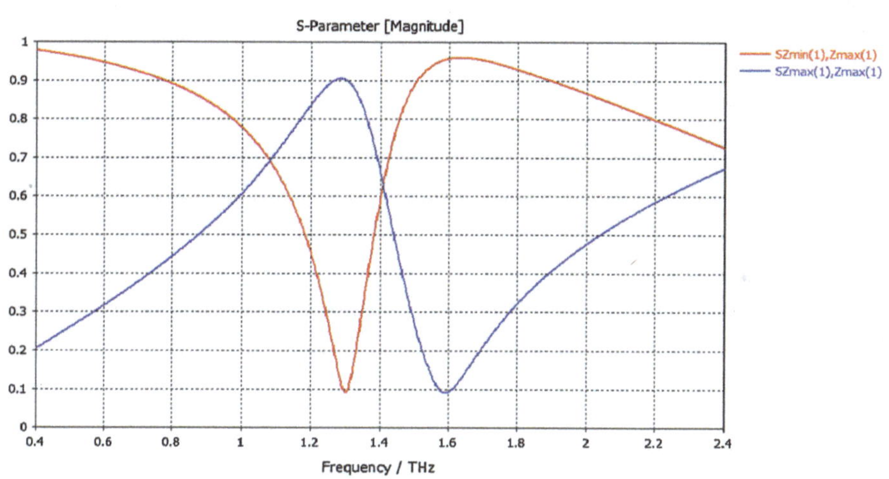

Fig. 5.31 Scattering parameters (S_{11}) and (S_{21}) characteristics of designed conformal RAM-type unit cell

Fig. 5.29a, b, respectively. From the simulation results, it is clear that designed conformal structure is resonating in THz regime as like planar one. The resonance frequency of the conformal structure is shifted due to the effect of curvature.

To verify the feasibility of the designed conformal unit cell as cloaking structure, the material properties are analyzed. The obtained scattering parameters for planar and conformal RAM structure are shown in Figs. 5.30 and 5.31. For the same, the

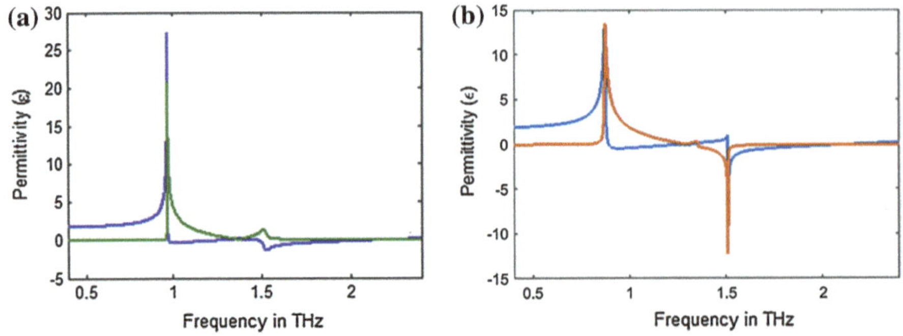

Fig. 5.32 Permittivity characteristics of proposed **a** planar RAM-type unit cell, **b** conformal RAM-type unit cell

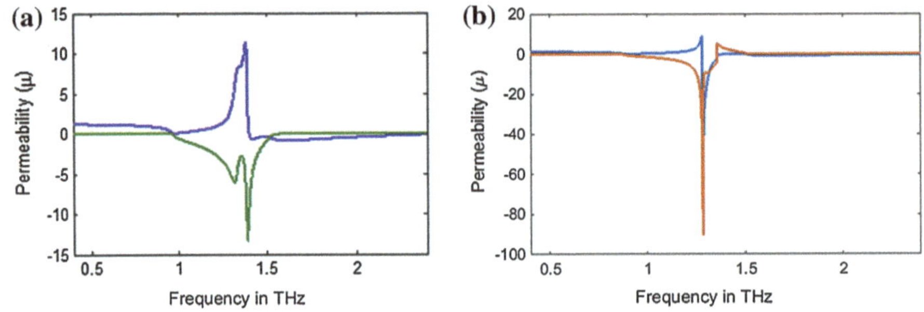

Fig. 5.33 Permeability characteristics of proposed **a** planar RAM-type unit cell, **b** conformal RAM-type unit cell

permittivity and permeability are retrieved from the S parameters using standard retrieval method. Figures 5.32 and 5.33 show the obtained material parameters along with planar one. From the graph, it is observed that conformal unit cell has much good negative characteristics as compared to the planar structure. So the designed conformal structure is suitable for cloaking applications.

5.8.4 Square SRR

The metamaterials are the materials which exhibit properties that are not naturally available. Split-ring resonators (SRR) are one of the most commonly used resonating metamaterial structures. In this design, square-shaped SRR with one square ring with a split in the middle is used. In this metamaterial structure, the middle split in the square ring acts as capacitor and results resonance at corresponding frequency. The unit cell dimensions are 50 μm × 50 μm × 2.5 μm. The square-shaped SRR has an outer dimension of 36 μm placed on the ultra-thin polyimide substrate. The line width is 0.2 μm and split width is 0.2 μm. The planar

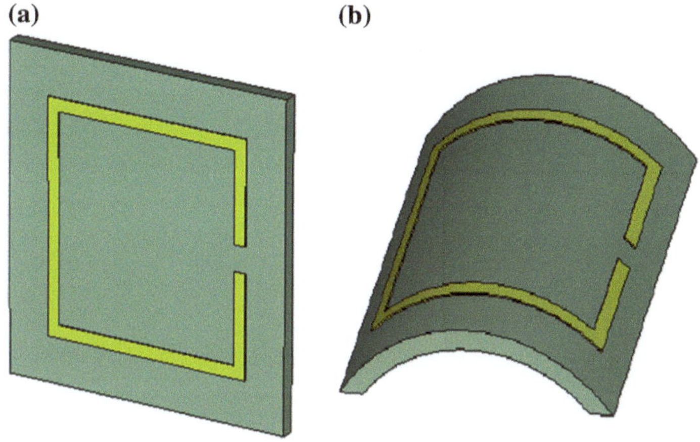

Fig. 5.34 **a** Typical proposed planar square-shaped unit cell, **b** typical designed conformal square-shaped unit cell

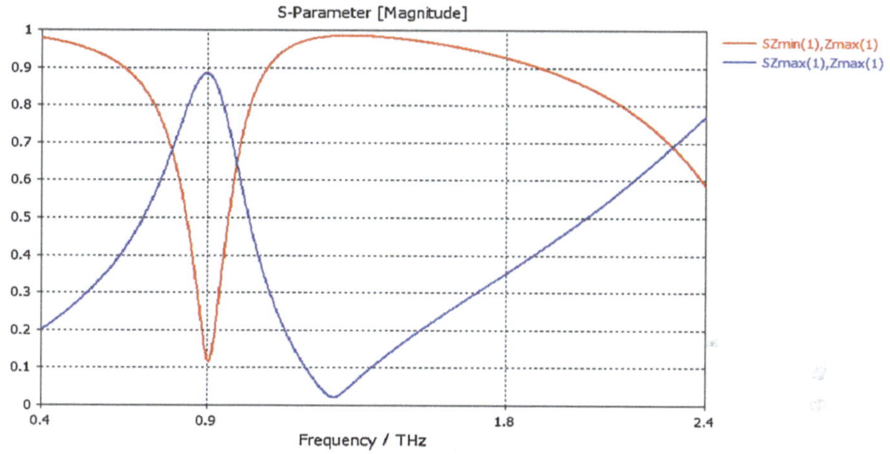

Fig. 5.35 Scattering parameters (S_{11}) and (S_{21}) characteristics of proposed planar square shape

structure converted to conformal by placing the designed metallic patch on the cylindrical sector which is etched out from the cylinder of radius 25 μm. Figure 5.34a, b represents the designed planar and conformal structure correspondingly.

FEM-based EM solver has been used to study the properties of conformed unit cell. The waveguide ports are defined, and excitations are given normally on the unit cell along the *z*-direction for obtaining the S parameters. The obtained reflection and transmission characteristics through simulation are shown in Figs. 5.35 and 5.36 along with the characteristics of corresponding planar structure.

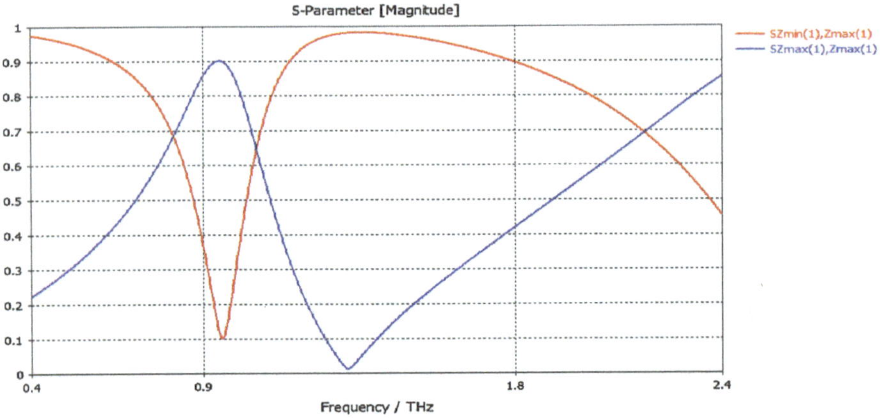

Fig. 5.36 Scattering parameters (S_{11}) and (S_{21}) characteristics of designed conformal square shape

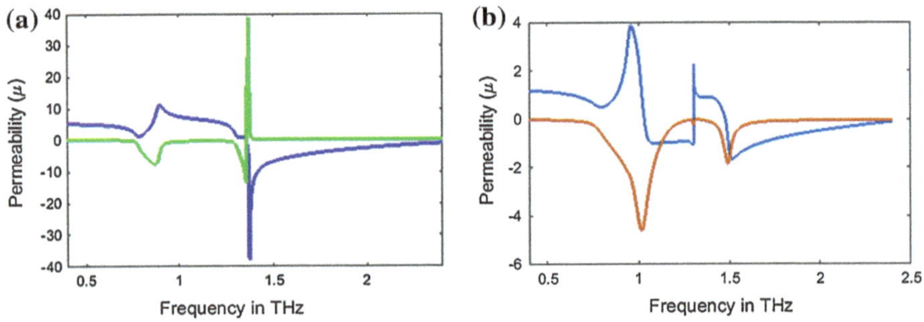

Fig. 5.37 **a** Permeability characteristics of proposed planar square-shaped unit cell, **b** permeability characteristics of designed conformal square-shaped unit cell

From simulated characteristics, it is clear that the characteristics of designed conformal structure are similar to proposed planar one. The simulation results indicate that the designed square SRR resonates at 0.9 THz with S_{11} of −30 db. The standard procedure is followed to extract the desirable permittivity and permeability properties from the return loss characteristics. The obtained material characteristics are plotted against the frequency and are shown in Figs. 5.37 and 5.38.

From the permittivity and permeability graphs, it is clear that the proposed structure exhibits negative permittivity and permeability characteristics in the frequency band 1–1.5 THz. It is observed that the negative characteristics of the designed conformal structure are inferior to the characteristics of proposed planar structure.

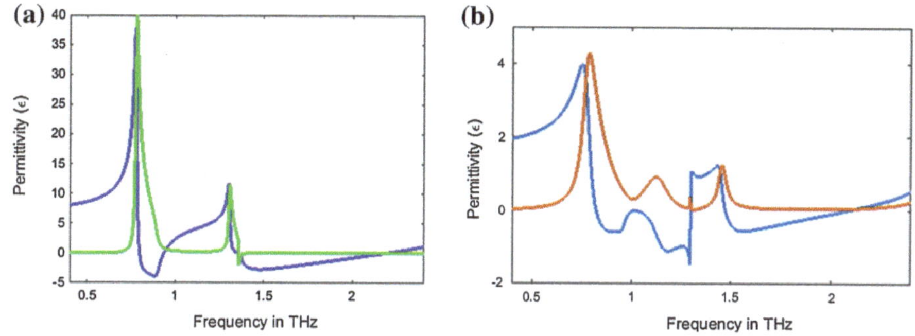

Fig. 5.38 Permittivity characteristics of proposed **a** planar square-shaped unit cell, **b** conformal square-shaped unit cell

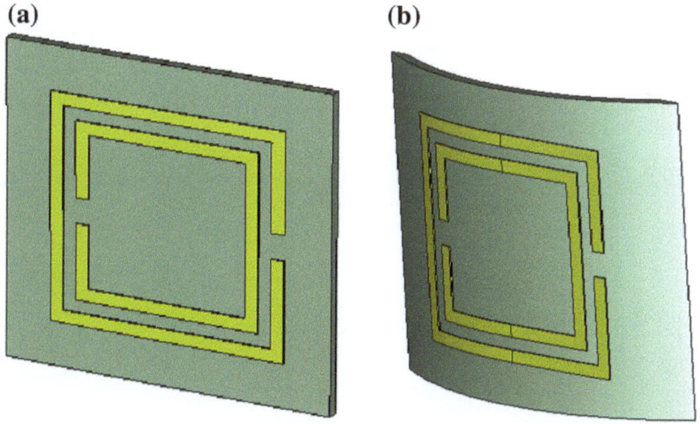

Fig. 5.39 Schematic of **a** planar double-square-shaped unit cell, **b** designed conformal double-square-shaped unit cell

5.8.5 Double-Square SRR with Gaps

The previous section already mentioned the design and analysis of square SRR. This section extends the study toward multiple ring structure for multiple resonance applications. The double-ring square SRR is designed using gold and is etched on polyimide substrate. The permittivity of substrate used is 3.5. The unit cell has a dimension of 50 μm × 50 μm × 1.5 μm, respectively. The structure is designed for terahertz frequencies. The size of the outer square SRR is 36 μm, with line width of 2 μm and the gap between the outer and inner rings is 2 μm. The size of split in the rings is 4 μm. The planar structure converted to conformal by placing the designed metallic patch on the cylindrical sector which is etched out from the cylinder of radius 25 μm. Figure 5.39a, b represents the designed planar and conformal double-ring SSRR structure, respectively.

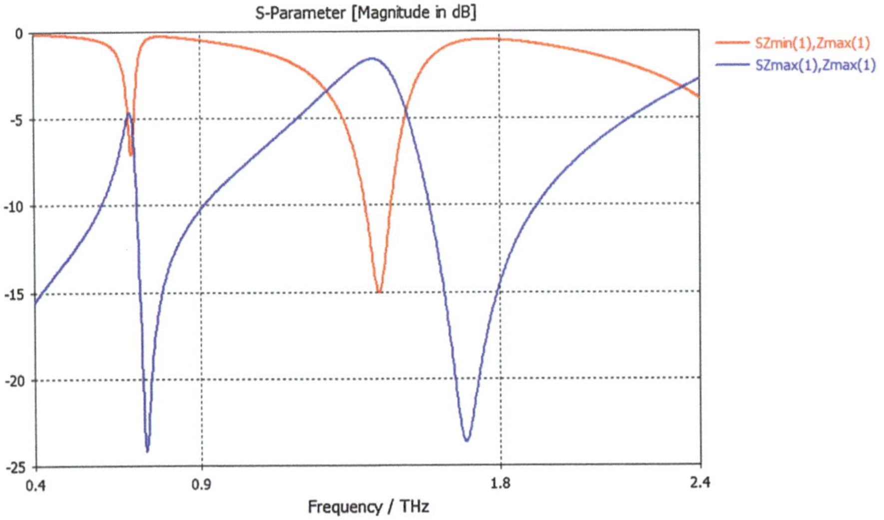

Fig. 5.40 Scattering parameters (S_{11}) and (S_{21}) characteristics of proposed planar double-square-shaped unit cell

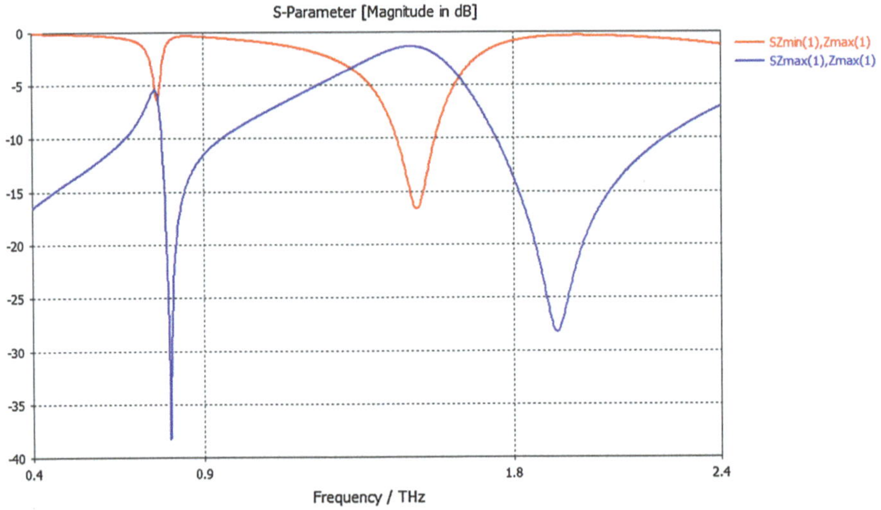

Fig. 5.41 Scattering parameters (S_{11}) and (S_{21}) characteristics of designed conformal double-square-shaped unit cell

FEM-based EM solver has been used to study the properties of conformed unit cell. The waveguide ports are defined, and excitations are given normally on the unit cell along the *z*-direction for obtaining the S parameters.

The obtained reflection and transmission characteristics through simulation are shown in Figs. 5.40 and 5.41 along with the characteristics of corresponding planar

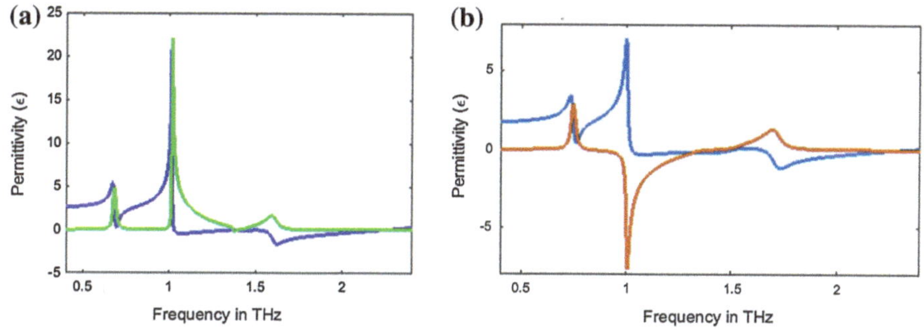

Fig. 5.42 Permittivity characteristics of **a** planar double-square-shaped unit cell, **b** conformal double-square-shaped unit cell

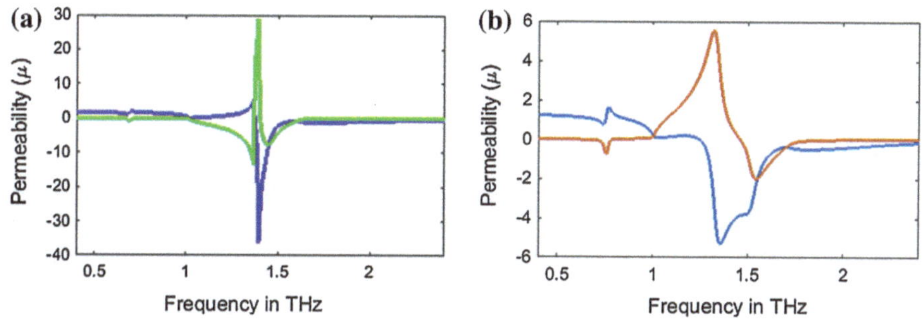

Fig. 5.43 Permeability characteristics of **a** planar double-square-shaped unit cell, **b** conformal double-square-shaped unit cell

structure. Scattering parameters show that multiple resonances are obtained using the designed structures and the resonance frequencies depend on the structural parameters of the inner and outer rings.

The material properties of designed conformal double-ring SSRR structure have been retrieved from obtained scattering parameters using standard S parameter retrieval method through MATLAB script. Figures 5.42 and 5.43 show the obtained material parameters along with planar one. The graph shows that the conformal unit cell exhibits negative characteristics, but which are not similar to characteristics of the planar structure.

5.8.6 *Middle Line SRR*

This section details the design of a conformal middle line SRR structure for cloaking applications. The structure consists of two cascaded SRR with a common middle line as shown in Fig. 5.44. A typical unit cell structure has dimension of

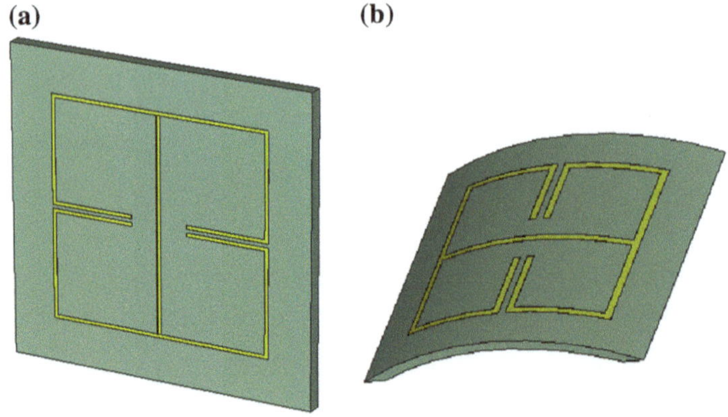

Fig. 5.44 **a** Typical proposed planar middle line SRR unit cell, **b** conformal middle line SRR unit cell

250 μm × 250 μm × 8.5 μm, respectively. The SRR has a longer dimension of 180 μm, line width of 2 μm and is made of gold. The dielectric constant of substrate used is 3.5.

The planar structure converted to conformal by placing the designed metallic patch on the cylindrical sector which is etched out from the cylinder, in order to study the feasibility of the structure for cloaking applications. Figure 5.44a, b represents the designed planar and conformal double-ring SSRR structures, respectively.

In order to characterize the permittivity and permeability of the proposed structure, the middle line SRR is simulated using an FEM-based EM solver. The corresponding return loss and transmission characteristic results are shown in Figs. 5.45 and 5.46. From the return loss curve, it is shown that multiple resonances occurred. These results pave the way for creating numerous multilayered non-planar electromagnetic composites that can lead to achieve better electromagnetic cloak.

The simulation results indicate that designed conformal unit cell structure has better scattering characteristics as compared to characteristics planar unit cell. The material properties of designed conformal middle line SRR structure have been retrieved from using S parameter retrieval method through MATLAB script. The obtained material parameters along with planar one are shown in Figs. 5.47 and 5.48. The graph shows that the conformal unit cell exhibits negative characteristics, but the characteristics are distinct from the planar ones.

5.8.7 Open Split-Ring Resonator

Split-ring resonators are one of the basic metamaterial structures which exhibits negative permeability and permittivity. This section introduces the design of a

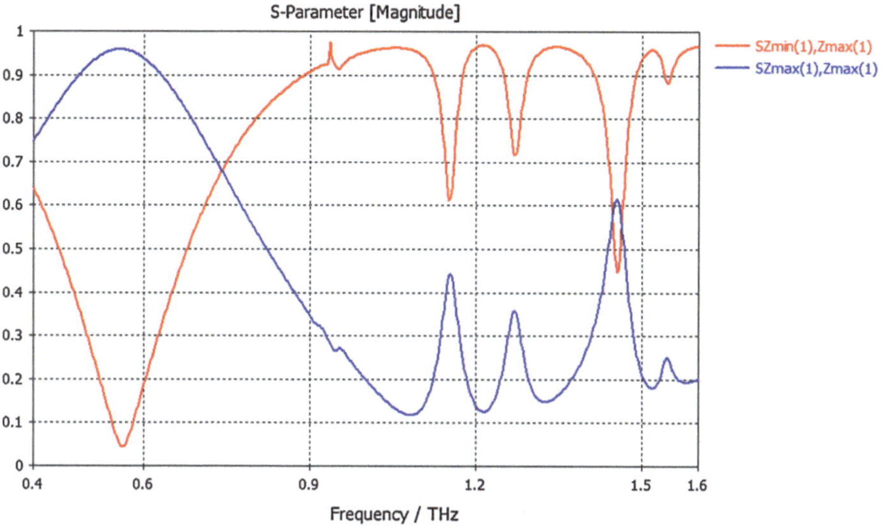

Fig. 5.45 Scattering parameters (S_{11}) and (S_{21}) characteristics of proposed planar middle line SRR

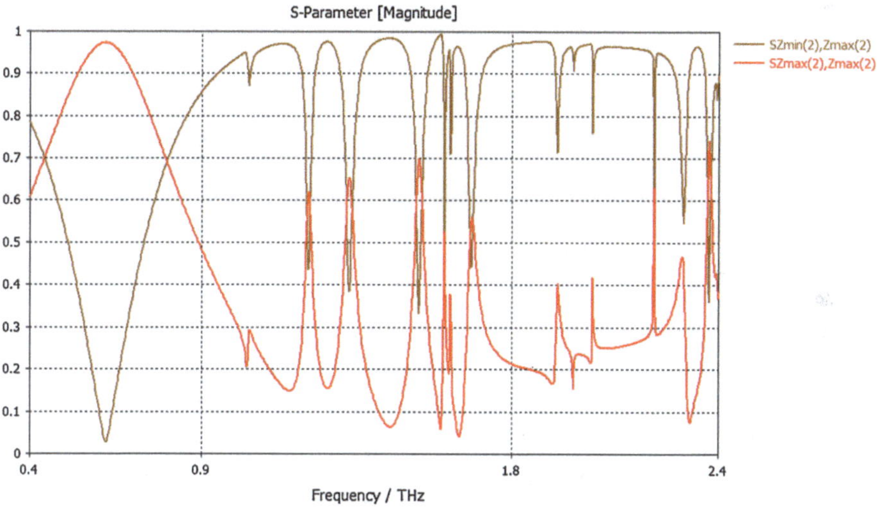

Fig. 5.46 Scattering parameters (S_{11}) and (S_{21}) characteristics of designed conformal middle line SRR

modified SRR called conformal open split-ring resonator. The main advantage of this proposed structure over regular SRR is the size miniaturization. The structure is formed by two concentric copper rings having openings at the same side and is connected to another set of concentric copper rings. The structure is printed on a cylindrical sector made of Duroid substrate as shown in Fig. 5.49. The length, width,

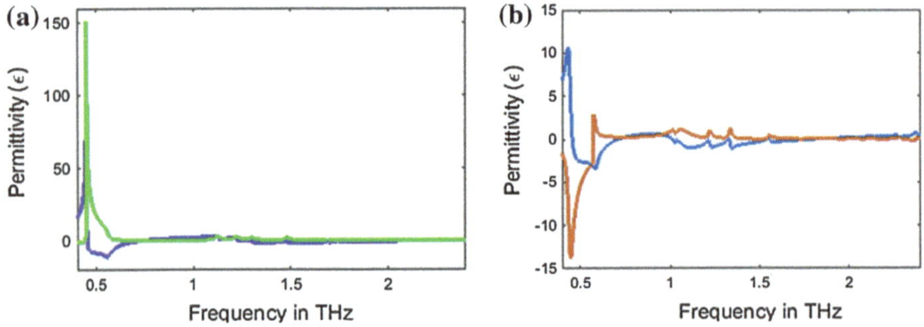

Fig. 5.47 **a** Permittivity characteristics of proposed middle line SRR, **b** permittivity characteristics of designed conformal middle line SRR

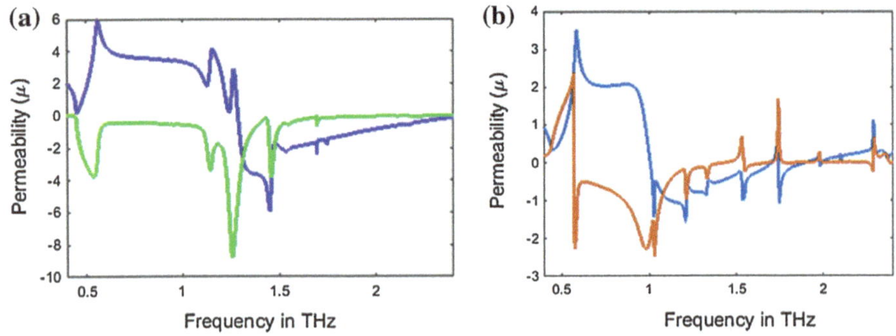

Fig. 5.48 Permeability characteristics of **a** planar middle line SRR, **b** conformal middle line SRR

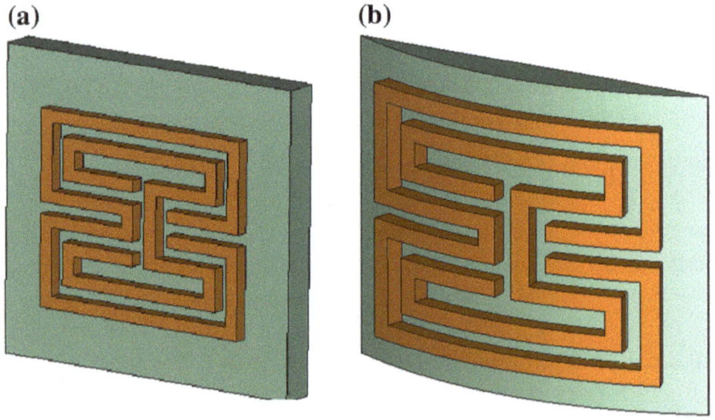

Fig. 5.49 **a** Typical proposed planar OSRR unit cell, **b** typical designed conformal OSRR unit cell

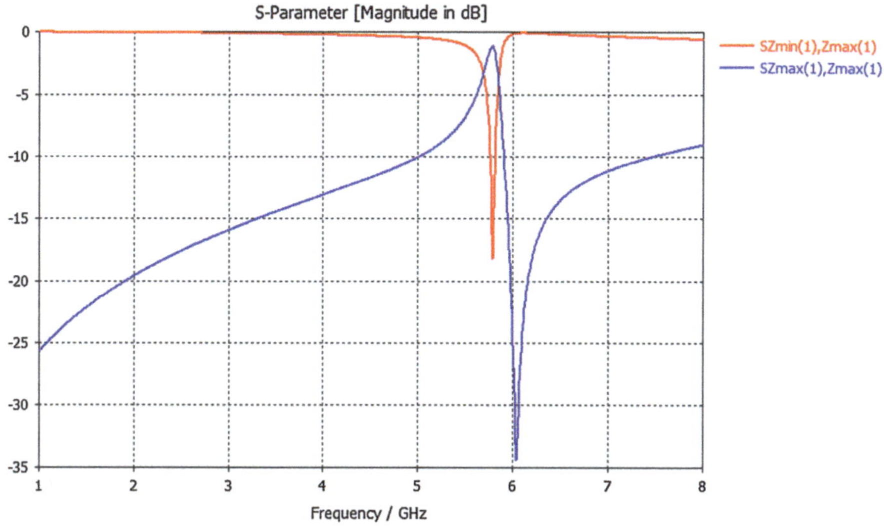

Fig. 5.50 Scattering parameters (S_{11}) and (S_{21}) characteristics of proposed planar OSRR

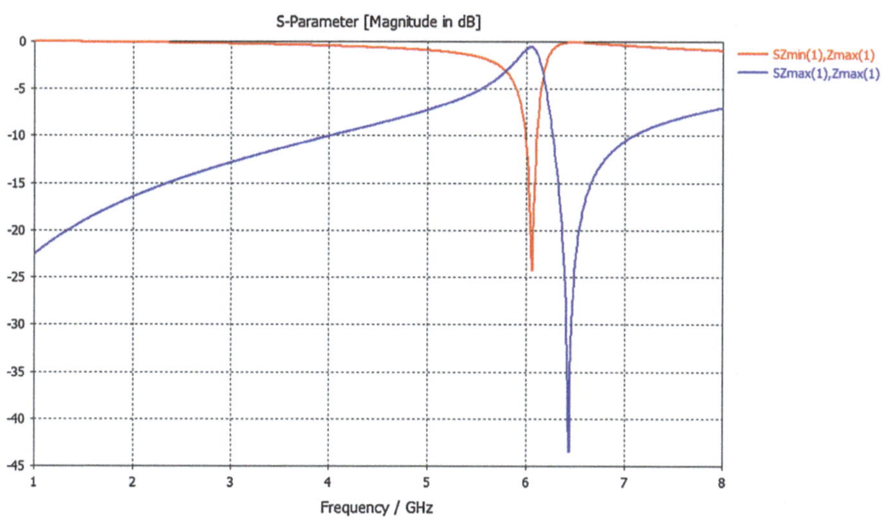

Fig. 5.51 Scattering parameters (S_{11}) and (S_{21}) characteristics of designed conformal OSRR

and height of the unit cell structure are 10 mm × 10 mm × 0.5 mm, respectively. The dielectric constant of a substrate is 2.33. The outer dimension in the SRR is 3.5 mm with line width of 1.5 mm. The thickness of the copper used is 0.2 mm.

FEM-based EM solver is used to design and simulate the proposed structure. This structure is placed inside a waveguide environment. The waveguide ports are assigned along the z-direction. The obtained S parameters through simulation are shown in Figs. 5.50 and 5.51 along with the characteristics of corresponding planar

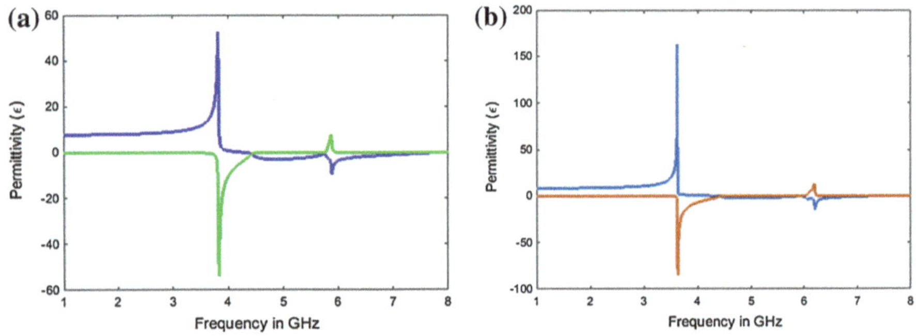

Fig. 5.52 **a** Permittivity characteristics of proposed planar OSRR, **b** permittivity characteristics of designed conformal OSRR

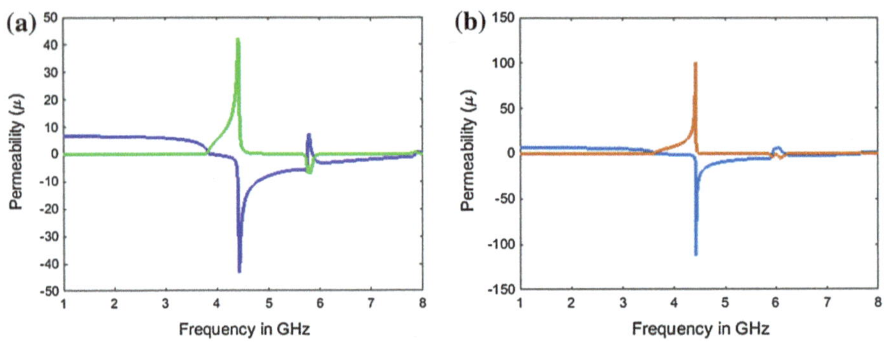

Fig. 5.53 Permeability characteristics of proposed **a** planar OSRR, **b** conformal OSRR

structure. From the simulated characteristics, it is clear that due to the effect of curvature, the conformal structure has better return loss characteristics as compared to the planar structure.

The permittivity and permeability values of the designed structure are retrieved from the S parameters and plotted as shown in Figs. 5.52 and 5.53. From the graph, it is clear that designed conformal open SRR shows negative characteristics better than planar one in the frequency range 4–6 GHz. Therefore, the OSRR can be used as double-negative (DNG) metamaterial cell.

5.8.8 Circular SRR

A conformal circular split-ring resonator (CSRR) with negative effective permeability is designed and compared with corresponding planar structure. The designed CSRR

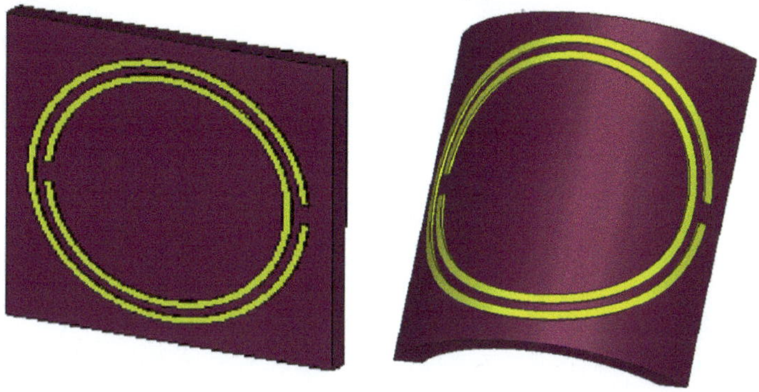

Fig. 5.54 **a** Typical proposed planar CSRR unit cell, **b** typical designed conformal CSRR unit cell

comprises of two circular rings with splits. The splits in outer and inner rings are identical in dimension and lie on the same axis towards opposite directions. The relative position of the gaps in the two rings affects the resonant frequency of CSRR. The structure is designed for L band operation, and the unit cell has a dimension of 12 mm × 12 mm × 0.8 mm, respectively. The planar CSRR structure is shown in Fig. 5.54a. In order to make the structure conformal, a cylinder of radius 6 mm is designed and a part of the desired substrate dimension is etched out. The metallic patch having same dimensions as that of planar one is conformed onto the cylindrical sector as shown in Fig. 5.54b. The metallic circular ring patches are made of gold and the bottom layer substrate having dielectric constant of 3.5. The designed conformal unit cell has been simulated using FEM-based simulation software.

The design and simulation of a planar and conformal structure has been carried out using FEM-based EM simulation software and return loss and transmission characteristics are shown in Figs. 5.55 and 5.56, respectively. The simulation results show that designed conformal structure is resonating as like planar one. Due to the curvature effect, the resonant frequency of the designed conformal structure was found to be shifted and showed less transmission when compared to that of planar one.

The material properties of designed conformal structure have been retrieved from obtained S parameters using standard retrieval method. Figures 5.57 and 5.58 show the obtained material parameters along with planar one. The graph shows that the conformal unit cell exhibits better negative characteristics as compared to the planar structure. So the designed conformal structure is suitable for cloaking applications.

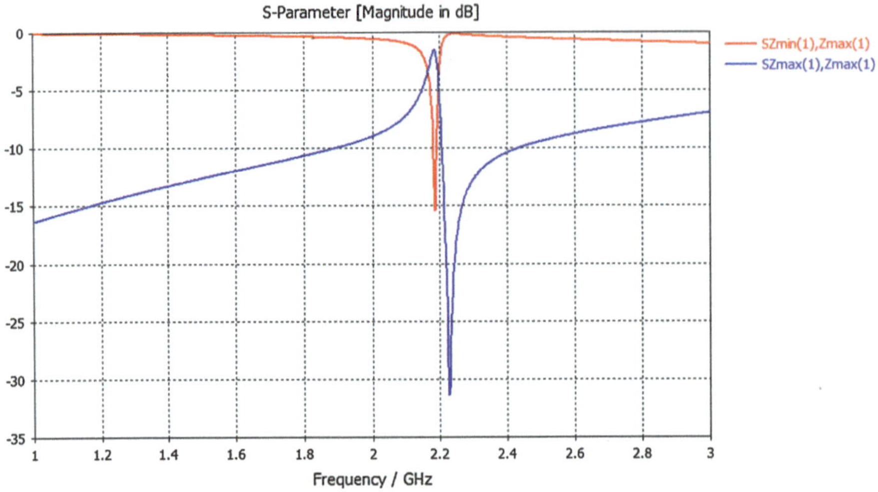

Fig. 5.55 Scattering parameters (S_{11}) and (S_{21}) characteristics of proposed planar CSRR

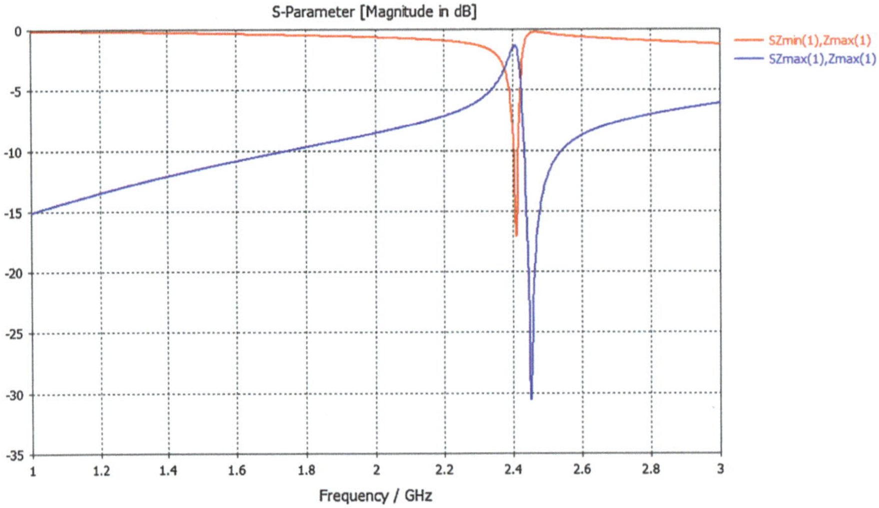

Fig. 5.56 Scattering parameters (S_{11}) and (S_{21}) characteristics of proposed conformal CSRR

5.8.9 Jerusalem Cross-Type Metamaterial Unit Cell

Jerusalem cross is one of the widely used metamaterial unit cell structure in the design of frequency selective surfaces (FSS). This section details the design and analysis of a planar and conformal Jerusalem cross and compares the characteristics of both structures. Further, the analysis has been extended toward the feasibility

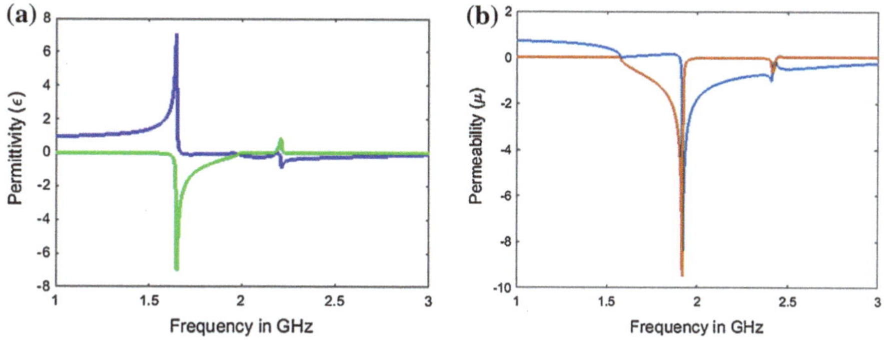

Fig. 5.57 Permittivity characteristics of proposed **a** planar CSRR, **b** conformal CSRR

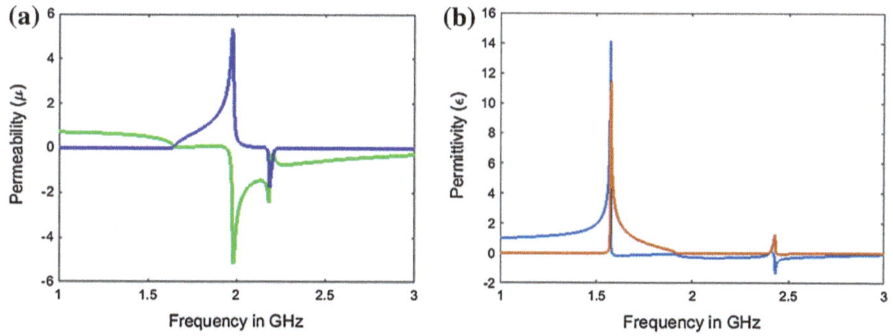

Fig. 5.58 Permeability characteristics of **a** planar CSRR, **b** designed conformal CSRR

study of the designed conformal structure in cloaking applications. Figure 5.59a shows the schematic of the proposed planar structure. The unit cell dimensions are 5 mm × 5 mm × 0.5 mm, respectively.

The Jerusalem-shaped metallic structure is made of copper having longer dimension of 3.6 mm and line width of 0.3 mm and is embedded on a substrate having dielectric constant of 2.55. In order to make conformal, a cylinder of radius 2.5 mm is designed and a sector of the desired substrate dimension is etched out. The metallic Jerusalem-shaped unit cell patch is conformed on to the cylindrical sector as shown in Fig. 5.59b.

The simulations of designed conformal structure are carried out through a FEM-based simulation software. Figures 5.60 and 5.61 show the reflection and transmission characteristics of proposed planar and designed conformal structure, respectively.

From the simulation results, it is clear that the scattering parameters show the same trend as in the case of planar one. The main advantage of the Jerusalem cross unit cell

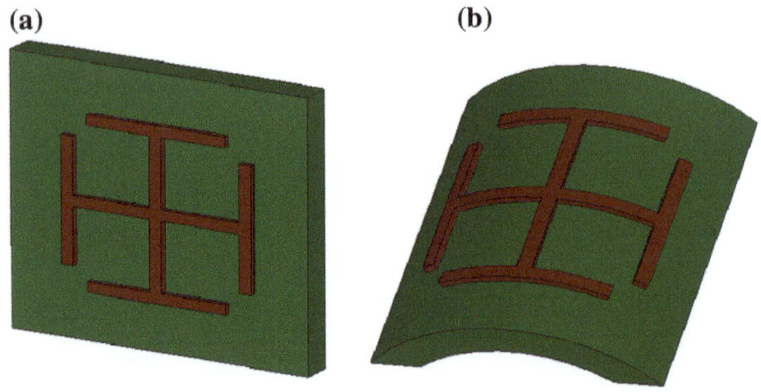

Fig. 5.59 Schematic of proposed **a** planar Jerusalem cross unit cell, **b** conformal Jerusalem cross unit cell

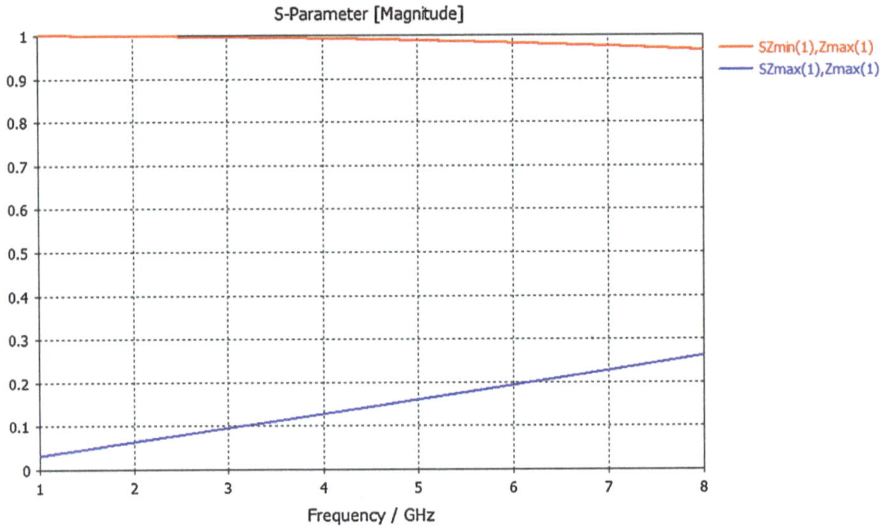

Fig. 5.60 Scattering parameters (S_{11}) and (S_{21}) characteristics of proposed planar Jerusalem cross unit cell

over other structures is that it exhibits wide angle and polarization-independent scattering characteristics due to the symmetry of the structure.

The material properties of a cloaking shell play an important role, in order to direct the electromagnetic wave. These complex material properties such as permittivity and permeability have been extracted from simulated scattering parameters of the metamaterial unit cells using a MATLAB script. The corresponding graphs are shown in Figs. 5.62 and 5.63.

It is observed that the negative characteristics of the designed conformal structure are inferior to the characteristics of proposed planar structure.

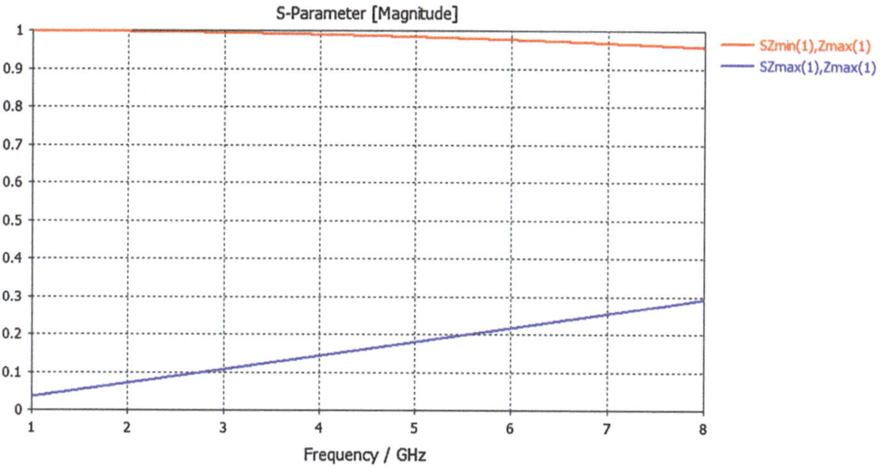

Fig. 5.61 Scattering parameters (S_{11}) and (S_{21}) characteristics of designed conformal Jerusalem cross unit cell

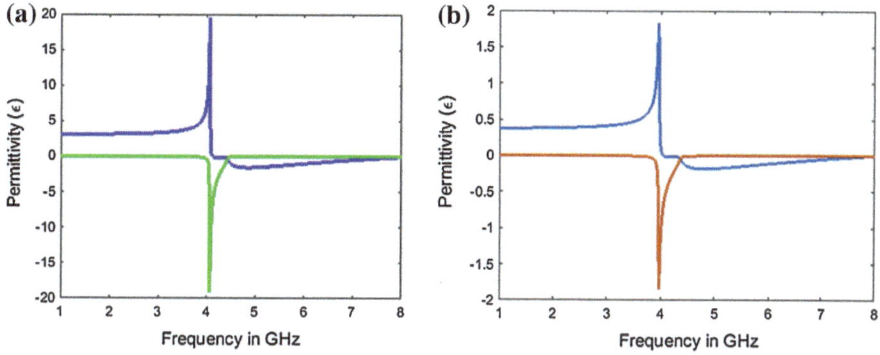

Fig. 5.62 Permittivity characteristics of proposed **a** planar Jerusalem cross unit cell, **b** conformal Jerusalem cross unit cell

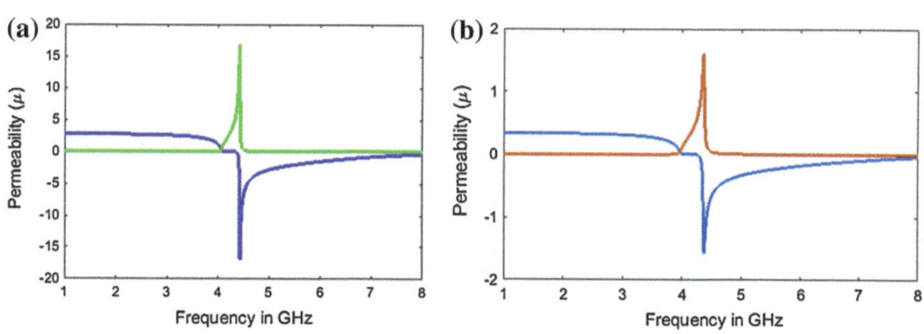

Fig. 5.63 Permittivity characteristics of proposed **a** planar Jerusalem cross unit cell, **b** conformal Jerusalem cross unit cell

5.9 Effect of Curvature

The concept of EM cloaking is nothing but hiding an object from electromagnetic wave incident on it. It founds extensive applications in modern military aviation, as stealth aircrafts are required to hide from enemies' radar capabilities in order to complete the mission in hostile territory. While the aerospace domain mostly deals with curved structures, it leads to the requirement of conformal cloaking unit cells. In this project, various conformal cloaking unit cells are proposed along with its planar corresponding. From the results, it has been observed that the curvature of unit cell affects the unit cell material properties. In this section, an attempt has been made to analyze the effect of curvature through designing a conformal I-shaped structure with different curvatures. Further, the analysis has been extended toward the optimization of the unit cell curvature for enhanced material properties. The design and simulation of conformal I-shaped structure for different curvatures are discussed in the next section.

5.9.1 Conformal I-Shaped Structure with Different Curvatures

The design and analysis of single conformal I-shaped structure is already explained in the previous section in detail. An array of I-shaped structure with different radius of curvatures has been designed and simulated using FEM-based EM simulation software. Figure 5.64 represents the schematic view of proposed design, which consists of three I-shaped structures arranged in two different curvatures.

The simulated reflection and transmission characteristics of the proposed structure are shown in Figs. 5.65 and 5.66. From the results, it is clear that as radius of curvature reduces, the scattering parameters are enhanced and thereby results in better negative material properties such as permittivity and permeability.

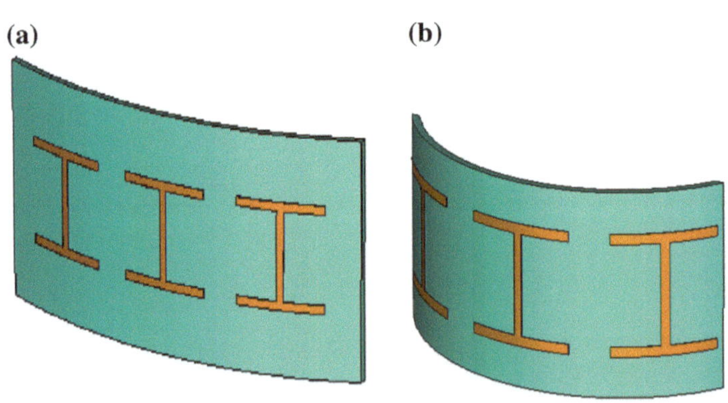

Fig. 5.64 Array of I-shaped unit cells conformed on **a** big curvature, **b** small curvature

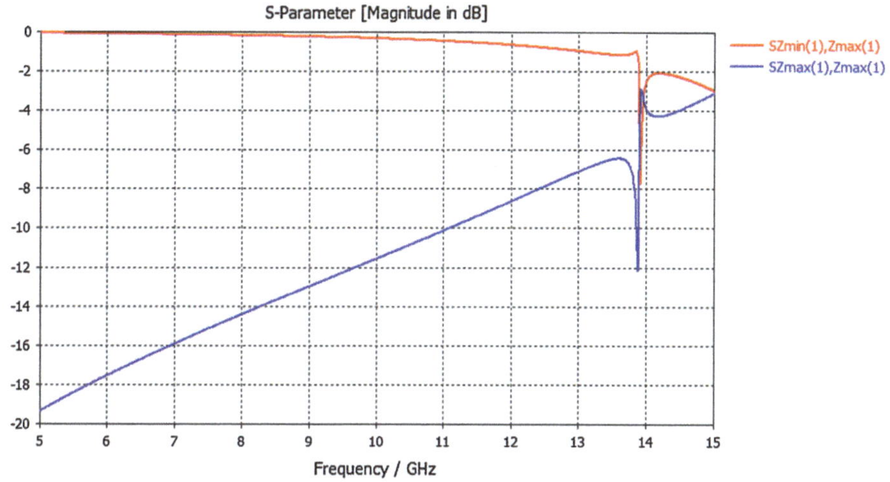

Fig. 5.65 Scattering parameters (S_{11}) and (S_{21}) characteristics of conformal I-shaped unit cell of big curvature

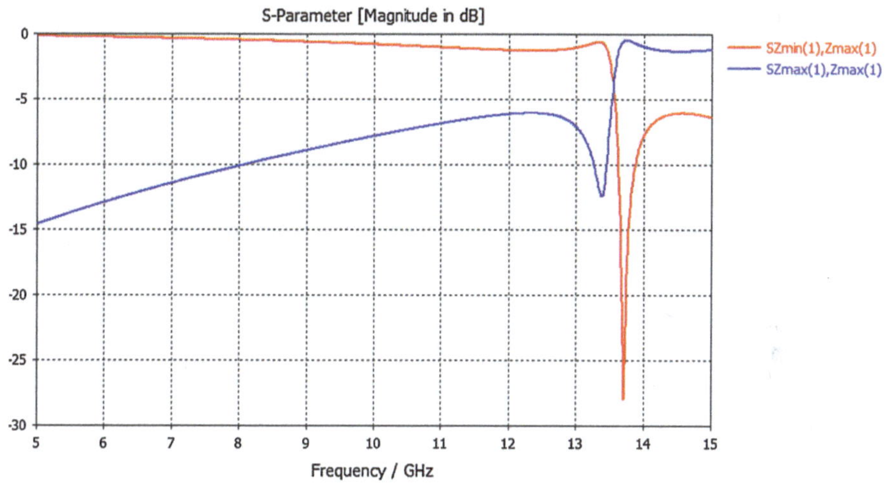

Fig. 5.66 Scattering parameters (S_{11}) and (S_{21}) characteristics of conformal I-shaped unit cell of small curvature

Standard S parameter retrieval method has been used to extract the material properties from the reflection and transmission characteristics obtained. The obtained permittivity and permeability characteristics are shown in Figs. 5.67 and 5.68, and it is crystal clear that the negative characteristics of permittivity and permeability are improved as radius of curvature reduces.

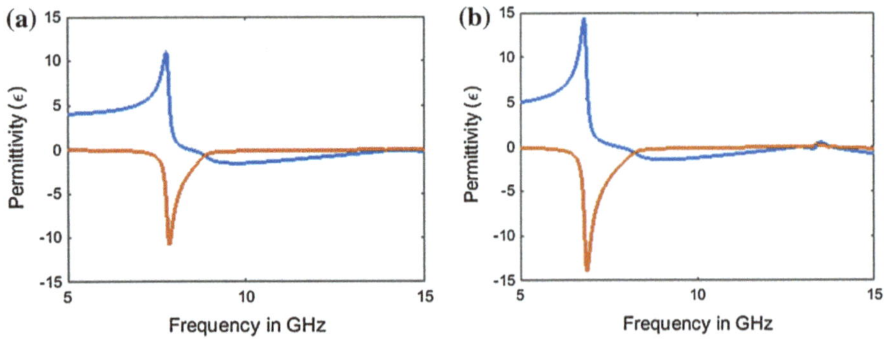

Fig. 5.67 Permittivity characteristics of conformal I-shaped unit cells with **a** big curvature, **b** small curvature

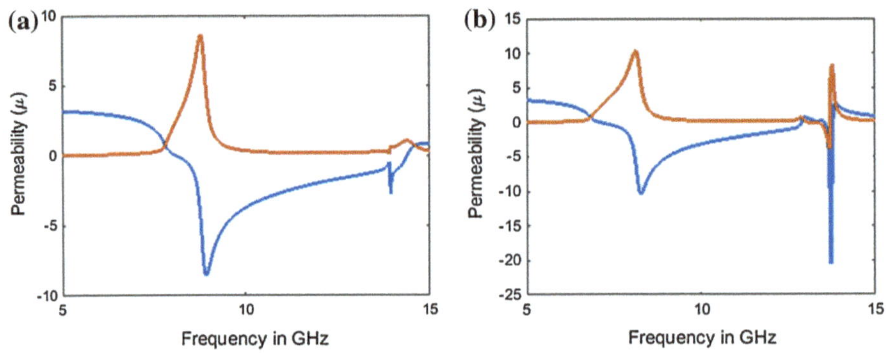

Fig. 5.68 Permeability characteristics of conformal I-shaped unit cells with **a** big curvature, **b** small curvature

The curvature of unit cell is optimized for optimum negative material properties using soft computing technique. The soft computing technique used is particle swarm optimization (PSO) algorithm. Soft computing techniques are one best option among different optimization techniques due to its unique features such as global optimization accuracy and quick convergence. The visual basic for applications scripting language is used to interface the optimization code with FEM-based solver.

5.10 Conclusion

Most of the cloaking applications require conformal cloaking shells. In this chapter, an effort has been made to design conformal metamaterial unit cell structures, which are suitable to use in the surface of curved cloaking shell. This chapter reveals the design of different conformal metamaterial unit cell structures from their planar corresponding. The designed structures are simulated using FEM-based EM solver, and the obtained scattering parameters of the conformal unit cells are compared with the corresponding planar structures. Further, in order to investigate the feasibility of the conformal structure as cloaking cell, the analysis of effective material parameters such as permittivity and permeability has been carried out. From the analysis, all the designed conformal structures exhibit negative material characteristics and are suitable for cloaking applications.

References

1. Leonhardt, U., and T. G. Philbin, "General relativity in electrical engineering," *New Journal of Physics,* vol. 8, pp., 1–18, 2006.
2. Leonhardt, U., and T. Tyc, "Broadband invisibility by non-Euclidean cloaking," *Science,* vol. 323, pp., 1–3, 2009.
3. Ma, H., "Material parameter equation for elliptical cylindrical cloaks," *Science,* vol. 312, pp., 1–4, 2008.
4. Tao, H., A.C. Strikwerda, X. Zhang, W. Adilla, C. M. Bingham, K. Fan, "Terahertz metamaterials on free-standing highly-flexible polyimide substrates," *Journal of Physics,* vol. 41, pp., 1–4, 2008.
5. Wanyi, G., H. Lianxing, S. Hao, Z. Hongwei, B. Li, S. Xiao-Wei, "A Dual-Band terahertz metamaterial based on A hybrid H-shaped cell," *Progress In Electromagnetics Research,* vol. 30, pp., 39–50, 2013.
6. Zhong, L., M. Cui, "Experimental realization of a broadband bend structure using gradient index metamaterials" *Optics Express,* vol.17, no. 20, pp., 18354–18362, 2009.

Chapter 6
Design Optimization of Broadband Radar Absorbing Structures

Anusha Eldo and Balamati Choudhury

6.1 Introduction

Stealth technology is not at all a new concept. As a matter of fact, stealth technology is totally the principle of absorption and reflection that makes aircraft "stealthy." Stealth technology also called low observable (LO) technology is the technique to make friendly aircrafts and warships less visible to the enemy's radar and detection capabilities in order to enhance its survivability. This is achieved either by deflecting the incident radar waves into some other directions or by absorbing the incident electromagnetic (EM) wave and hence the number of waves which returns to the radar is reduced.

Radar cross-section (RCS) reduction is a prominent field of research for achieving stealth in military domain. RCS is the measure of back-scattered electromagnetic waves from the target as it is illuminated by an incident wave. The detectability of the target is measured in terms of RCS reduction and hence the enhancement in target survivability.

Generally, the radar cross-section reduction techniques or stealth techniques fall into one of the four categories [1]: target shaping, materials selection and coating, passive cancellation, and active cancellation. By these techniques, the incident EM signal is cancelled or absorbed by the absorber coating. Also, out of these techniques, none of them alone reduces the RCS in effect. The conventional methods of radar absorbers give RCS reduction in narrow frequency range. But modern military applications demand for broadband frequency of operation, hence EM absorption in broad frequency range is required.

The introduction of artificially engineered materials turned out to be a possible solution for achieving broadband EM absorption. The artificially engineered materials

A. Eldo (✉) · B. Choudhury
Centre for Electromagnetics, CSIR-National Aerospace Laboratories,
Old Airport Road, Bengaluru 560017, India
e-mail: anushaeldho27@gmail.com

B. Choudhury (ed.), *Metamaterial Inspired Electromagnetic Applications*,
DOI 10.1007/978-981-10-3836-5_6

such as metamaterials are engineered to endow with useful modern electromagnetic devices such as EM absorbers, super lenses, and EM cloaks [2, 3]. These techniques turn out to be possible due to their special properties such as negative refractive index or resonance upshot and the crystal effects such as electronic band gap. In recent years, the near-perfect absorption characteristics of metamaterial-based structures have been broadly investigated. The metamaterial absorbers are made of periodic sub-wavelength resonators and are considered as homogeneous medium. The absorption capability of metamaterial-based structures is entirely depending on thickness and other structural parameters of the periodic patterns. To achieve optimum design, multi-objective particle swarm optimization-based computational engine has been used. The optimization through computational engine provides optimized design parameters of broadband absorbers.

In this chapter, bandwidth-enhanced metamaterial absorbers for microwave frequency regime have been designed and fabricated. The broadband metamaterial absorbers comprise of special resonant structures, with multiple resonances in the near frequencies. The optimum design for the structure has been achieved through soft-computing-based computational engine. The numerical results obtained from simulations shows that the pentagon structure exhibits more than 90% absorption for a bandwidth of 2.3 GHz and novel structure provides near-perfect absorption ranging from 8 to 11 GHz. Further the analysis of electric field distribution and surface current distribution is carried out, in order to understand the resonance mechanism of the structure. The experimental results for fabricated test samples also ensure the near-perfect absorption for wide band frequencies.

6.2 Radar Absorber Design Procedure

The detection of a target is represented in terms of radar cross section and is the measure of a target's ability to reflect radar signal in the direction of radar receiver. It is the measure of backscatter power per steradian in the direction of radar to the power density that is intercepted by the target. The RCS can be expressed [4] as

$$\sigma = \underset{a \to \infty}{\mathrm{lt}} 4\pi a^2 \frac{|E^{\mathrm{s}}|^2}{|E^{\mathrm{i}}|^2} \tag{6.1}$$

where a is the distance from radar, E^{s} indicates the scattered electric field and E^{i} is the incident electric field at the target.

The basic concepts used for radar absorbing structure design are matched characteristic impedance and wave impedance. The former represents the characteristic impedance at the surface of structure which is equal to the free space impedance, ensuring zero reflection; latter is the concept where the input impedance which is equal to free space impedance ensures complete transmission of energy into the absorber. The impedance of RAM is given by

$$Z = \sqrt{\frac{\mu}{\varepsilon}} = \sqrt{\frac{\mu_r}{\varepsilon_r}} \tag{6.2}$$

For perfect matching, the ratio of relative permittivity to relative permeability must be unity. The imaginary parts of permeability and permittivity of a material represent the losses associated with that. The relation between absorption, reflection, and transmission is given by

$$|T^2| + |R^2| + A = 1 \tag{6.3}$$

when $T = R = 0$, absorption become unity.

6.3 Background Theory on Design and Optimization

The lack of naturally available materials for EM absorption leads to the exploration of artificially engineered materials. The design of artificially engineered materials like metamaterials is a complicated procedure and requires the aid of commercially available EM solvers for design verification. As stealth platform in aerospace domain demands for reduced thickness of radar absorbing structure, optimization of various structural parameters became necessary. This section gives an idea about the design and optimization techniques used for metamaterial-based RAS structure.

6.3.1 Metamaterial-Based RAS

Metamaterials are the materials which exhibits certain properties that are not seen in naturally occurring materials, the most exciting one being the ability to show negative refractive index. The word metamaterial was proposed by Rodger M. Wasler, University of Texas. The root of this word, "meta," is Greek and it means beyond—i.e., metamaterial means beyond the material. Though the science of metamaterials is relatively new, Victor Veselago predicted the possibility of existence of doubly negative materials in 1968, in a Russian publication [5, 6].

Smith et al. (2000) fabricated the negative index (NI) material by using the metamaterial structures [7]. The proposed metamaterial unit cell consists of combination of a wire structure with negative permittivity and a split-ring resonator with negative permeability for the same frequency band, resulting negative refraction.

Shelby et al. (2001) demonstrated the confirmation of negative refraction, through Snell's law experiment by using metamaterial wedge in prism shape [8]. The incident microwave radiation on the surface of prism was refracted to the opposite side of normal to the surface and hence shows the negative refraction.

Jain (2006) explained the fabrication of electromagnetic absorbers, radio frequency lenses, and phantom materials by using the artificial dielectrics. He also developed the formulae for permittivity estimation of artificial dielectrics [9].

Pang et al. (2012) proposed a broadband and ultra-thin high-impedance absorber on a metamaterial substrate for microwave frequencies [10]. The proposed structure consists of split-ring resonators (SRRs) embedded vertically into dielectric slab. This structure gives an expanded bandwidth of absorption for less than −10 dB. The field distribution characteristics showed that the enhanced characteristics related to the LC resonance. The results demonstrate the suitability of metamaterials along with high-impedance surface absorber for bandwidth broadening.

6.3.2 Broadband Radar Absorbing Structures

Modern military applications always demand for broadband absorbers. A number of literatures are already available regarding the broadband, wide angle, and polarization-independent radar absorbers. Some of them are explored below.

Yang and Shen (2007) presented the design of a broadband absorber by using double square loops as unit cells [11]. The introduced design contains lumped resistors of suitable value in all the four sides of the loop. These loops are printed on a dielectric substrate with a conducting or metallic back plate, in order to minimize the transmission. The structure was simulated in HFSS and results showed that 10 dB RCS reduction is achieved over a range of 8–18 GHz frequency with relative bandwidth of 77%.

Wakatsuchi et al. (2010) presented a broadband metamaterial absorber for arbitrary polarization by using lossy cut wire (CW) pair metamaterials [12]. The lossy cut wire pairs with different lengths are combined as a periodic unit and are placed near to perfect electric conductor (PEC). The simulations are performed by using transmission line modelling. The main advantage of lossy cut wires as compared to other metamaterial structures are simplicity of structure and no need of any extra metal component to yield Negative Refractive Index (NRI). The paired CW exhibits both electric resonance and magnetic resonance, also it works for both polarizations. The structure gives an absorption peak of 76% at 27 GHz.

Grant et al. (2011) demonstrated design, simulation, fabrication, and measurement of a broadband metamaterial absorber in terahertz frequency range [13]. The closely positioned resonant peaks of stacked metal-insulator layers with different structural parameters made the absorption spectrum broader. For a bandwidth of 1.86 THz, the obtained absorption for the proposed structure was more than 60%. Also the FWHM (full width at half maximum) bandwidth obtained was about 48%, and it found application in bolometric terahertz imaging.

Liu et al. (2012) designed and fabricated a metamaterial absorber with broadband absorption at microwave frequencies [14]. The designed structure consists of circular metallic patches, PEC ground plane and dielectric spacer. For broadband resonance characteristics, a number of metallic patches with adjacent resonant peaks are used. Experimental results showed that the structure has an absorption bandwidth of 2.8 GHz and relative FWHM bandwidth of 25.3%.

Yuan and Cheng (2014) designed and experimentally validated lumped element-based metamaterial absorber for broadband absorption in the microwave frequency range [15]. In this composite metamaterial absorber (CMMA), the EM wave energy is converted to electrical energy and subsequently consumed by lumped resistors. The designed CMMA has absorption characteristics insensitive to polarization and angle of incidence. The experimental results showed that CMMA yields below −10 dB in the frequency band from 2.85 GHz to 5.31 GHz with relative bandwidth of 60.3%. So metamaterials are one of the mainly used artificially engineered materials for radar absorbing applications.

6.3.3 Optimization of RAS Using PSO

As the absorption characteristics depend on physical dimensions of the structure, it is required to find out parameters for best performance. But as mathematical formulations for designs are not readily available, this is an iterative and cumbersome task. As a result, a demand for a technique that can identify the optimum dimensions arises. Further, need to minimize the total thickness of the absorber arises in order to meet design requirements. Multi-objective particle swarm optimization (MOPSO) has proven to be an effective tool for design optimizations [16].

The particle swarm optimization for RAM/RAS design is not a new concept. In fact, researchers demonstrated the implementation of modified versions of the same for designing RAMs. This improved PSO was faster and at the same time prevented the algorithm from falling into local minima.

Multi-objective pareto-front-based algorithms are used to achieve optimization of more than one objective function simultaneously. Chamaani et al. (2007) [17] used a modified MOPSO technique in order to optimize a five-layered RAM.

Cui and Weile (2005) [18] proposed the implementation of a parallel PSO for RAM design. The fitness function is calculated for each agent in parallel and updates the global best, after calculating the fitness of all the agents only. This type of PSO is called synchronous PSO. In contrast, conventional or asynchronous PSO updates the global best after calculating the fitness of each agent. For some type of problems, this technique was found to be faster than asynchronous PSO and Genetic algorithm.

In the coming section, the theory behind multi-objective particle swarm optimization is discussed followed by the development of MOPSO computational engine. This computational engine will be used to design an optimized radar absorbing structure.

6.4 MOPSO for Design Optimization of RAS

High-absorption performance and minimum thickness are two requirements encountered during design of absorber. These two design requirements are conflicting with each other, and designer is forced to arrive at a trade-off between these two design objectives. Manually trying to obtain solution that is best favorable to both these conditions is nearly impossible. Therefore, advanced multi-objective optimization strategies like multi-objective particle swarm optimization must be used to arrive at a set of solutions wherein one value of fitness cannot be reduced without severely increasing the others.

6.4.1 Development of Computational Engine for Optimization of Radar Absorber

While the theory behind the working of absorbers is well known, mathematical formulation for the design of metamaterial-based absorbers is not available. Equations for the design of metamaterial itself may be found; however, the presence of additional dielectrics in the construction of absorbers shifts the resonant frequency. This lack of mathematical formulation necessitates the integration of the optimization tool with EM simulators in order to achieve optimization through iterative EM simulation. In this work, the soft computing technique used is particle swarm optimization algorithm. The reason for using this algorithm is explained in the previous section. The PSO algorithm is implemented in MATLAB. The PSO code acts like a kernel, which calls a finite integration method (FIT)-based EM solver in order to determine the fitness as shown in Fig. 6.1. This technique involves calling the solver-specific commands such as visual basic language (VBA) in the MATLAB framework. The PSO algorithm, along with the EM solver, creates a computational engine for optimization.

The developed computation engine passes the coordinates of each particle during every step in the iteration to the EM solver. Using this dimensional information, and solver-specific commands, which involves assigning of material properties to each component in the structure, the solver is able to design the structure, assign boundary conditions, assign EM solver parameters, simulate the structure for a particular frequency range, and output the S-parameters obtained from the simulation. These S-parameters are sent back to the PSO kernel, which uses these values in order to calculate an application specific fitness function. This iterative procedure returns the optimized structural parameters for the specific fitness function.

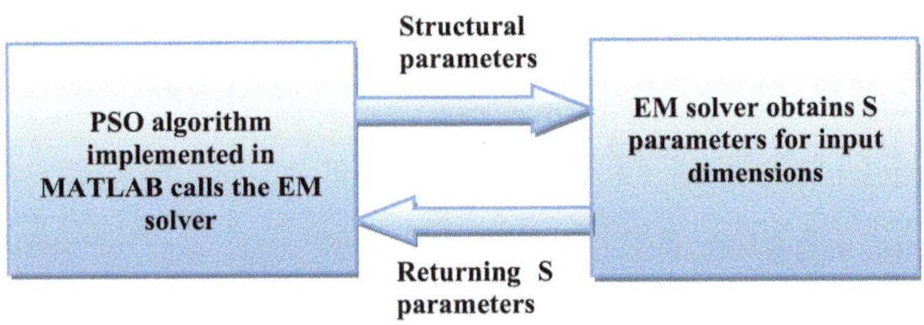

Fig. 6.1 Schematic of the PSO-based computational engine

6.5 Design Methodology and Simulation Results

In this chapter two metamaterial-based radar absorbing structures for broadband absorption are designed and simulated using FEM-based EM simulation software. For the optimization of structural parameters, PSO algorithm with fitness function as absorption coefficient is used. In this section, the design methodology and analysis of the results of both pentagons-shaped structure and novel metamaterial-based RAS structure have been explained.

6.5.1 Design of Pentagon Structure

Metamaterial absorber consists of periodic sub-wavelength resonating structures, which are capable of exhibiting near-perfect absorption performance. The periodic metal resonator provides the electric response to tune the permittivity of the combined structure, whereas the ground plane tunes the magnetic permeability of the structure. As the effective impedance of the structure satisfies the free space impedance, the reflection became minimum value. Also as compared to conventional materials, the thickness of the structure is much less and design is more efficient and flexible.

Wang et al. (2014) proposed an ultra-thin metamaterial structure for broadband performance [19]. The proposed structure is pentagon-shaped and consists of two dual-band sub-cells with some air gap between the two layers. While optimizing the various structural parameters of the proposed structure through soft computing technique PSO, better absorption results are obtained.

Design of Unit Cell

The perspective view of the optimized pentagon-shaped metamaterial structure is shown in Fig. 6.2a. The metamaterial structure consists of three layers, metallic patterns, dielectric substrate, and metallic back plate. Metallic patterns determine the electric response, whereas thickness and material of the substrate layer are responsible for the magnetic behavior of the metamaterial structure. This facilitates the tuning of absorption characteristics through varying the thickness and type of dielectric used. The metallic inclusions in the structure are made of copper with frequency independent conductivity of $\sigma = 5.8 \times 10^7$ S/m. The dielectric substrate used is epoxy glass cloth laminate (FR4) with relative permittivity of 4.3 and loss tangent of 0.025 with thickness of 1.6 mm. The unit cell has a size of 10 mm × 10 mm. The geometric parameters of the structure are given by, side of pentagon A = 6 mm, distance to fifth side of pentagon from centre is d = 1.5 mm.

The electromagnetic absorption by an absorber can be formulated as, $A = 1 - R - T = 1 - |S_{11}|^2 - |S_{21}|^2$. As the thickness of the metallic back plate is more than skin depth corresponding to the frequency of interest, the transmission from the structure is negligible. So the absorption value is mostly depending on reflections from the structure. Reduced reflection from the surface of the structure

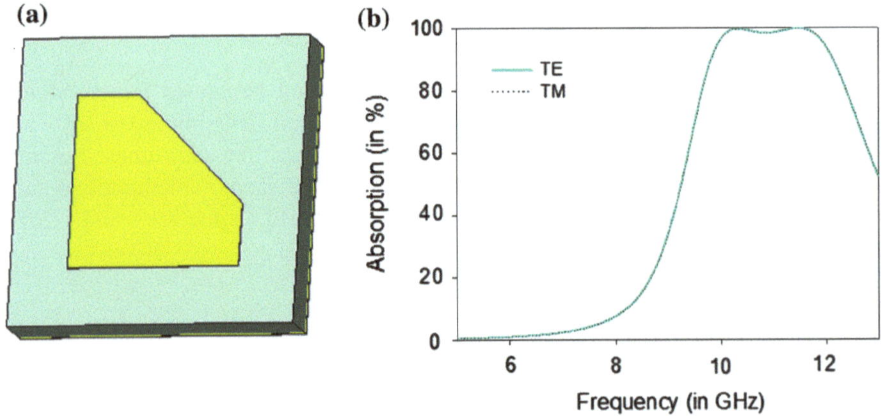

Fig. 6.2 **a** Schematic of proposed pentagon-shaped metamaterial structure, **b** reflection characteristics of the proposed structure

enhances the absorption efficiency. In contrast, the unit cell can also be considered as an RLC equivalent resonance circuit, which is tuned by varying the structural parameters and dielectric substrate used. At RLC resonance, perfect matching with free space impedance occurs, and energy is lost in the dielectric without any transmission.

The lack of mathematical formulations relating the absorber performance and geometrical parameters of the structure necessitates the integration of the optimization tool with EM simulators, in order to obtain optimum design. The optimum design is achieved through soft-computing-based optimization algorithms along with FEM-based EM simulation tools. The particle swarm optimization algorithm-based computational engine has been developed by interfacing the FEM-based simulation tool to the kernel MATLAB optimization code.

The reflection characteristics for proposed structure are obtained through FEM-based full wave electromagnetic solver. For the simulation of unit cell, periodic boundary conditions are applied along x and y directions, and plane wave is normally incident on the patterned absorber surface along $+z$ direction with electric field x-polarized and magnetic field y-polarized. The absorption calculated from reflectivity is shown in Fig. 6.2b. Two absorption peaks are obtained at $f1$ = 10.3 GHz and $f2$ = 11.48 GHz with absorption of 99.99%. The dual-band metamaterial absorber has continuous absorption spectrum ranging from 9.8 to 12 GHz with absorption above 90%. It reveals that the impedance offered by the metamaterial structure is tuned to match with the free space impedance in the absorption spectrum. Also, the proposed structure shows polarization-insensitive absorption characteristics and is depicted in Fig. 6.2b.

The analysis has been extended toward the field distribution and surface current flow, to get better understanding of the resonance mechanism and formation of absorption peaks. The surface current and field distribution are simulated using FEM-based solver and are shown in Figs. 6.3, 6.4 and 6.5. In the depicted figures,

(a) (b)

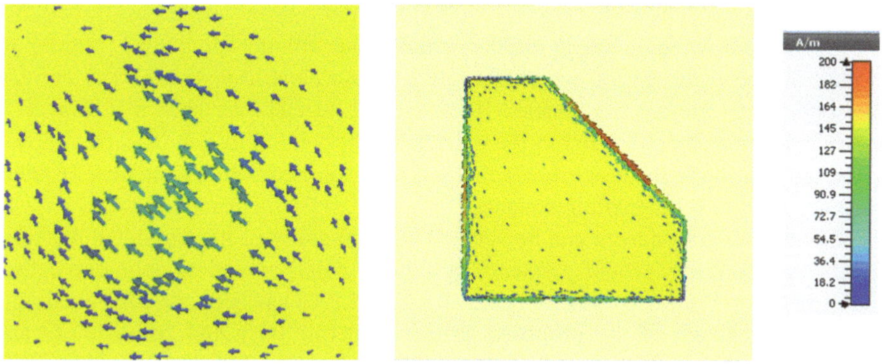

Fig. 6.3 Surface current distribution at resonant frequency 10.32 GHz in **a** *top layer* (pattern), **b** *bottom layer* (ground plane)

(a) (b)

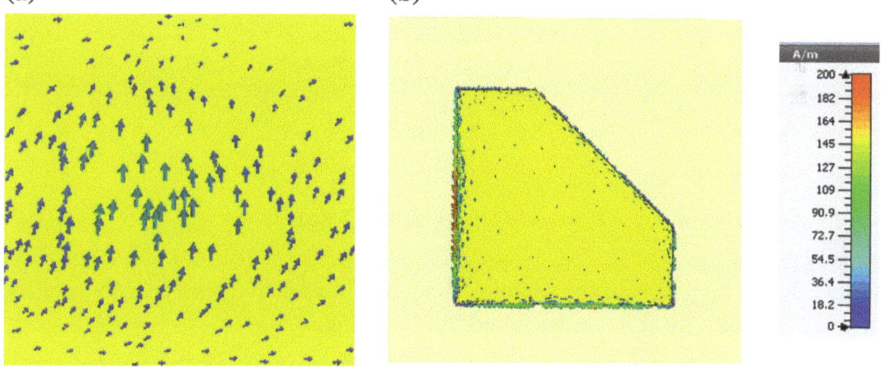

Fig. 6.4 Surface current distribution at resonant frequency 11.49 GHz in **a** *top layer* (pattern) **b** *bottom layer* (ground plane)

(a) (b)

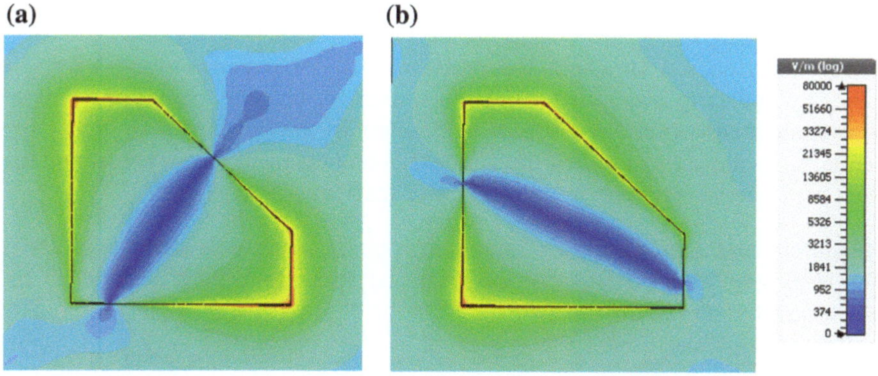

Fig. 6.5 Electric field distribution at resonant frequency of **a** 10.32 GHz, **b** 11.49 GHz

intensity of field is represented by color and current flow direction is indicated by arrows.

As the EM wave incident on the structure, a counter-circulating current is formed on the patch edges, which depicts a strong electric response. The current distribution at resonant frequencies 10.32 and 11.49 GHz are shown in Figs. 6.5 and 6.6. The current exhibited in the pattern is anti-parallel to that of current flows in the metallic back plate; it reveals that current loop is driven by incident H field, resulting in strong magnetic resonance.

The metamaterial design can be verified through study of its equivalent material parameters, permittivity and permeability. The popular extraction technique called Nicolson Ross–Weir S-parameter retrieval method [20] has been used for obtaining material parameters of the proposed structure. Permittivity and permeability are obtained from the reflection and transmission coefficients through equations.

$$\eta_s = \pm\eta_0\sqrt{\frac{(s_{11}+1)^2 - s_{21}^2}{(s_{11}-1)^2 - s_{21}^2}} \tag{6.4}$$

$$z = \frac{s_{21}(\eta_s + \eta_0)}{(\eta_s + \eta_0) - s_{11}(\eta_s - \eta_0)} \tag{6.5}$$

$$k_s = \frac{j}{a}\ln z \tag{6.6}$$

$$k_s = \frac{1}{a}[-\arg z + 2p\pi + j\,\ln|z|] \tag{6.7}$$

where a is the thickness of the sample. The value of effective permeability (μ_s) and effective permittivity (ε_s) can be found out by using

$$\varepsilon_s = \frac{k_s}{\omega\eta_s} \tag{6.8}$$

$$\mu_s = \frac{k_s\eta_s}{\omega} \tag{6.9}$$

For the proposed structure, the retrieved material parameters are shown in Fig. 6.6. The proposed structure exhibits negative relative permittivity in the absorption band width. For resonant metamaterials, the obtained permittivity and permeability are seen to follow the Drude–Lorentz characteristic. This feature has been used extensively in the optimization of metamaterial design.

Simulation and Analysis

The absorption characteristics of the proposed structure entirely depend on the dielectric properties of the substrate used and the dimensions of the metallic pattern. In this section, we discuss the impact of various parameters such as substrate

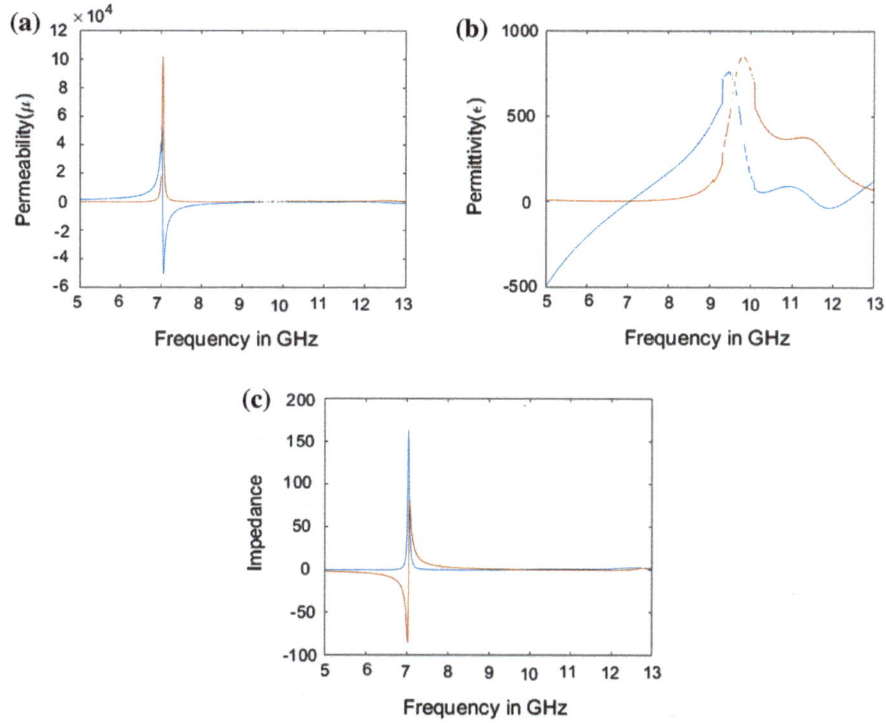

Fig. 6.6 Retrieved material parameters for the designed structure **a** relative permeability, **b** relative permittivity **c** effective impedance

thickness, periodicity, size of pentagon, angle of incidence, and polarization on absorption performance of proposed metamaterial structure.

Effect of Thickness of Dielectric (t) The thickness of dielectric layer has much effect on determining the resonance frequency as well as the absorption efficiency. The performance of the absorber for different dielectric thickness is as shown in Fig. 6.7. As thickness t increases from 1.4 to 1.8 mm, the dual-band resonance becomes single resonance peak. So the absorption bandwidth decreases. The absorption performance was improving as the thickness increases, and near-perfect absorption is attained for thickness more than 1.6 mm.

Effect of Periodicity In determining absorptive frequency spectrum, periodicity is one of the important factors. To analyze the effect of periodicity, p is varied from 8 to 12 mm and is illustrated in Fig. 6.8. As periodicity varies from 8 to 12 mm, the dual resonance peaks became one resonant peak and resonant peak is retained at 10.32 GHz. As the periodicity increases, the absorption efficiency decreases. So absorption bandwidth is enhanced at the cost of absorption efficiency.

Effect of Size of Pentagon Parametric study in Fig. 6.9 shows that size of the pentagon pattern (*A*) and distance to fifth side from center of pattern (d) have much

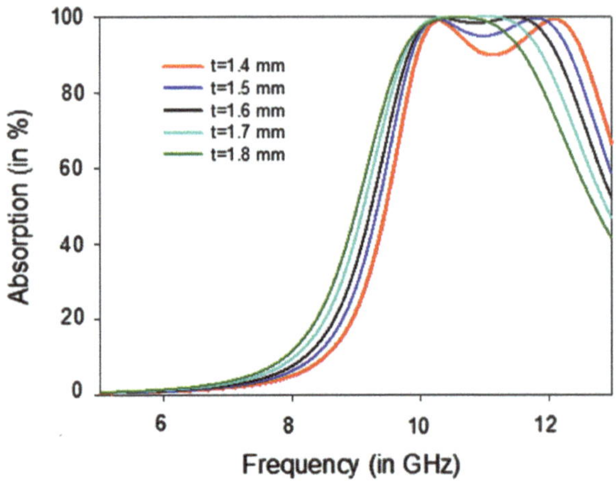

Fig. 6.7 Variation in absorption with thickness of dielectric changed from 1.4 to 1.8 mm

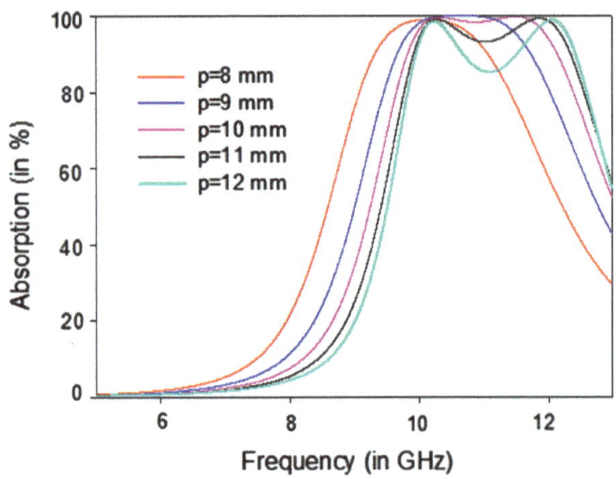

Fig. 6.8 Variation in absorption with periodicity of structure changed from 8 to 12 mm

important role in determining the frequency of absorption as well as bandwidth. The structure is optimized for maximum bandwidth for near-perfect absorption.

As size of pentagon increases from 5 to 7 mm, the frequency of operation shifts to left; however, the bandwidth remains almost same. While the distance to fifth side (*d*) increases, the absorption bandwidth reduced significantly, but the frequency of operation remains same. Through computational engine-based optimization, *A* is chosen as 6 mm and *d* as 1.5 mm for optimum performance. For this optimum design, the real part of impedance is equal to 377 Ω for a broad frequency range of 2.3 GHz, and the structure acts as a good absorbing material.

Effect of Angle of Incidence and Polarization The absorption property of the proposed structure is studied for different angle of incidence and polarization. Figure 6.10 shows the effect of different incident angles such as 0°, 20°, 40°, 60°, 80°, respectively. The absorption bandwidth remains constant up to 30° and

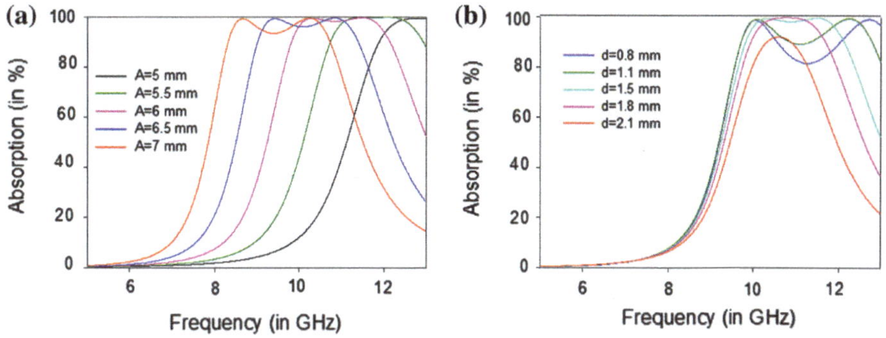

Fig. 6.9 Variation in absorption with **a** size of pentagon, **b** distance to fifth side from center of pentagon

Fig. 6.10 Variation in absorption for angle of incidence changed from 0° to 80° in steps of 20°

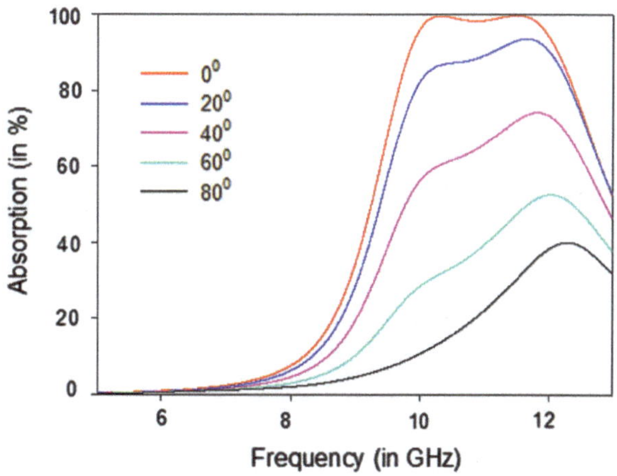

subsequently the absorption reduced. It is also to be observed that the structure loses the near-perfect absorption property when the angle of incidence is greater than 40°. The comparison regarding the effect of TM- and TE-polarized plane wave incidences is already shown in Fig. 6.2b. The proposed structure shows good polarization stability for TM and TE-polarized waves for small incident angles.

6.5.2 Design of Novel Metamaterial Structure

A novel metamaterial absorbing structure is designed, for broadband operation. The main disadvantage of conventional metamaterial structures is the narrow band of absorption. So an effort has been made to overcome this challenge in absorption bandwidth. In this chapter, we introduce a novel broadband metamaterial absorber with dual-band resonance. The schematic of the proposed metamaterial unit cell is

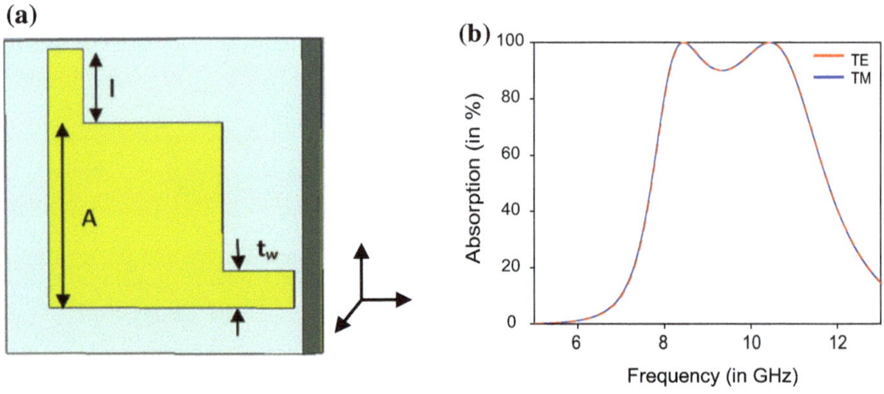

Fig. 6.11 **a** Unit cell structure for proposed metamaterial absorber. **b** Absorption characteristics for TE and TM polarizations

shown in Fig. 6.11, structure consist from top to bottom as metallic pattern which is electric resonator, dielectric substrate, and metallic back plate. The thickness and type of dielectric layer determines the magnetic response. So we are able to tune the absorption characteristics by varying the dimensions of pattern and thickness of the dielectric. All the metal used in the structure are made of copper with conductivity of $\sigma = 5.8 \times 10^7$ S/m (frequency independent conductivity). Dielectric substrate used is epoxy glass cloth laminate (FR4) with relative permittivity of 4.3 and loss tangent of 0.025 with a thickness of 1.6 mm. The geometric parameters of the structure is given by, sides of square $A = 5$ mm, length of tail $l = 1.7$ mm, and width of tail $t_w = 1$ mm.

The electromagnetic absorptive efficiency of an absorber is characterized by $A = 1 - R - T = 1 - |S_{11}|^2 - |S_{21}|^2$ where A is the absorbance, S_{11} and S_{21} are reflection and transmission coefficients, respectively. As the structure contains metallic back plate with thickness more than skin depth, the transmission value (S_{21}) for the structure is zero. Hence, the absorption entirely depends on the reflection from the structure. Absorption is maximized through minimization of reflections from the top of the structure. In contrast, the unit cell can also be considered as an RLC equivalent resonance circuit, which is tuned by varying the structural parameters and dielectric substrate used. At RLC resonance, perfect matching with free space impedance occurs and energy is lost at the dielectric without any transmission. The optimum geometrical dimensions are chosen by soft-computing-based computational engine.

The reflection characteristics for proposed structure is obtained through FEM-based full wave electromagnetic solver. For the simulation of unit cell, periodic boundary conditions are applied along x and y directions and plane wave is normally incident on the patterned absorber surface along $+z$ direction with electric field x-polarized and magnetic field y-polarized. The absorption calculated from reflectivity is shown in Fig. 6.11b. Two absorption peaks are obtained at $f_1 = 8.43$ GHz and $f_2 = 10.42$ GHz with absorption of 99.99%. The dual-band metamaterial absorber has continuous absorption spectrum ranging from 8 to 11 GHz with absorption above 90%. It reveals that impedance offered by metamaterial structure is tuned to match with free space impedance in the absorption spectrum. Also, the proposed structure shows polarization-insensitive absorption characteristics, and these characteristics are depicted in Fig. 6.11b.

The analysis has been extended toward the field distribution and surface current flow, to get better understanding of the resonance mechanism and formation of absorption peaks. The surface current and field distribution are simulated using FEM-based solver and are shown in Figs. 6.12, 6.13, and 6.14. In the depicted figures, intensity of field is represented by color and current flow direction is indicated by arrows.

The current exhibited in the pattern is anti-parallel to that of current flow in the metallic back plate, and it reveals that current loop is driven by incident H field, resulting strong magnetic resonance. The electric field distribution in the structure

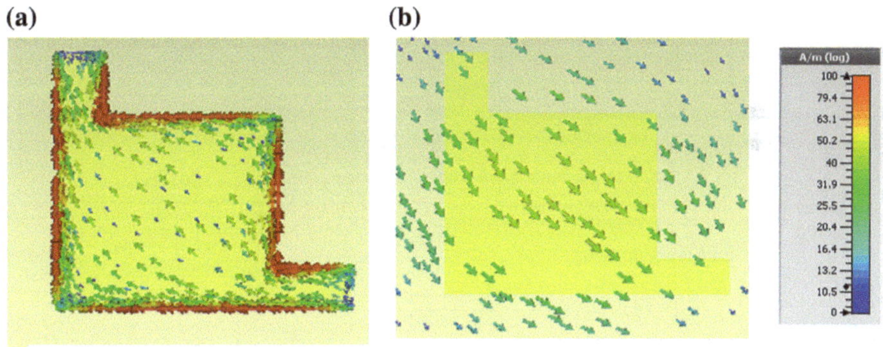

Fig. 6.12 Surface current distribution at resonant frequency 8.43 GHz in **a** *top layer* (pattern), **b** *bottom layer* (ground plane)

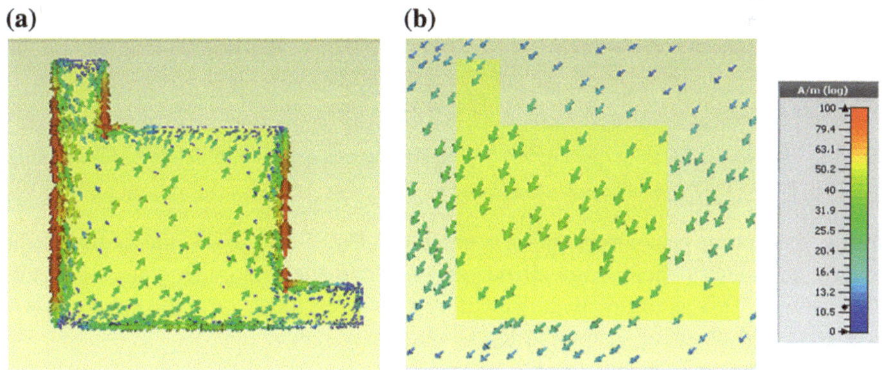

Fig. 6.13 Surface current distribution at resonant frequency 10.42 GHz in **a** *top layer* (pattern), **b** *bottom layer* (ground plane)

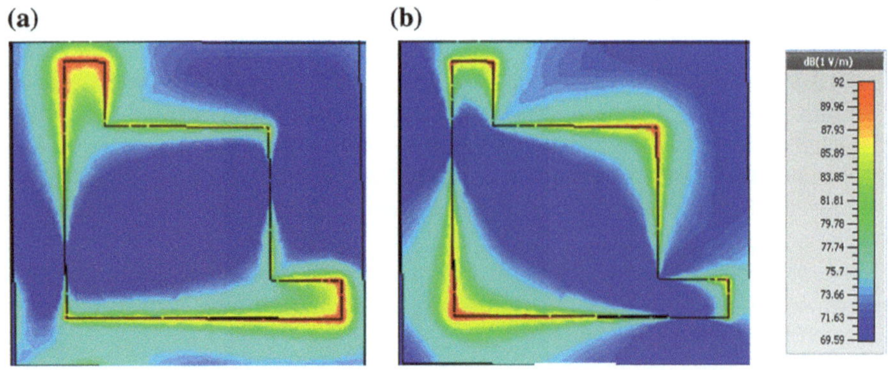

Fig. 6.14 Electric field distribution at resonant frequencies at **a** 8.43 GHz, **b** 10.42 GHz

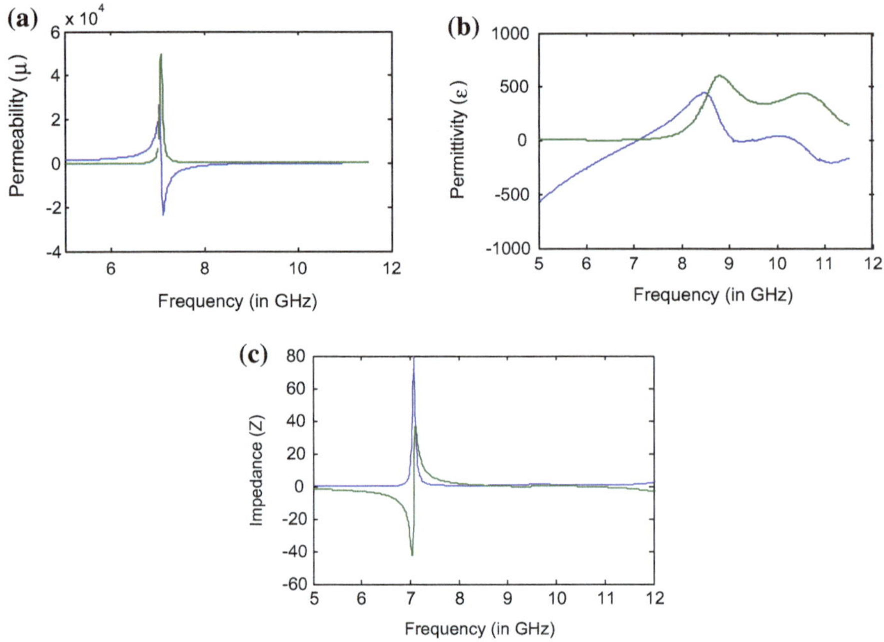

Fig. 6.15 Retrieved material parameters for the designed structure **a** relative permeability, **b** relative permittivity, **c** effective impedance

during absorption peak frequencies is shown in Fig. 6.14a, b. At the edges of the pattern, the electric field is localized, and most of the energy is lost due to this confinement. This strengthens the electromagnetic resonance, and the structure acts as a good EM absorber in the particular frequency range.

The metamaterial design can be verified through study of its equivalent material parameters, permittivity and permeability. For the proposed novel structure, the retrieved material parameters are shown in Fig. 6.15.

The proposed structure exhibits negative relative permittivity in the absorption bandwidth. For resonant metamaterials, the obtained permittivity and permeability are seen to follow a Drude–Lorentz characteristic. This feature has been used extensively in the optimization of metamaterial design.

Simulation and Analysis

The absorptive nature of the structure is entirely depending on dielectric used and dimensions of pattern. In this section, we discuss the impact of various parameters such as substrate thickness, periodicity, tail length, angle of incidence and polarization on absorption performance of proposed metamaterial structure.

Effect of Thickness of Dielectric The thickness of dielectric layer has much effect on determining resonance frequency as well as on absorption efficiency. The performance of the absorber for different dielectric thickness is as shown in Fig. 6.16. As thickness increases from 1.4 to 2.2 mm, the resonant frequency is shifted to left and also at thickness greater than 2 mm, only one resonant peak occurs. The absorption performance was improving as the thickness increases, and near-perfect absorption is attained for thickness more than 1.6 mm.

Effect of Periodicity In determining absorptive frequency spectrum, periodicity is one of the important factors. To analyze the effect of periodicity, p is varied from 8 to 11 mm and is illustrated in Fig. 6.17. As periodicity varies from 8 to 11 mm, the first resonant peak f_1 is shifted from 8.2 to 8.85 GHz and second resonance peak f_2

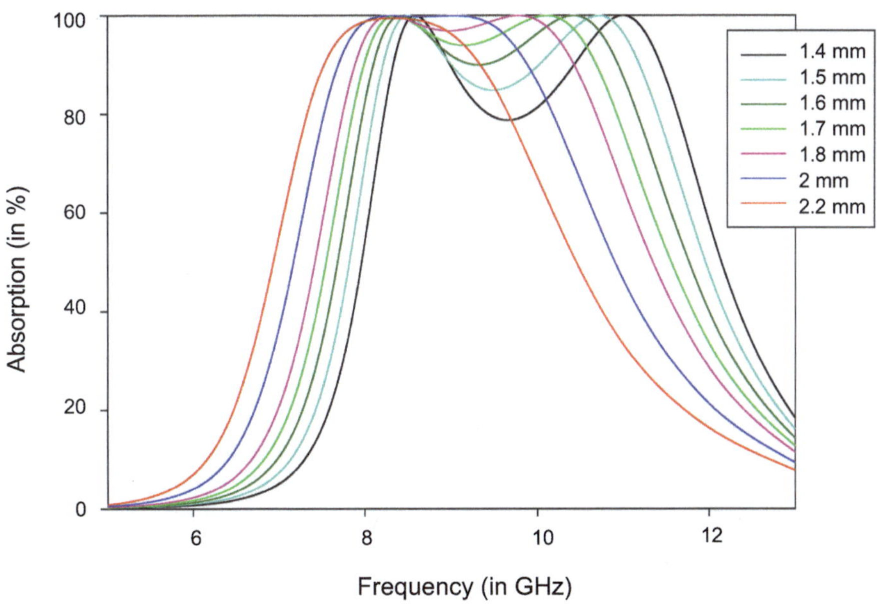

Fig. 6.16 Variation in absorption with thickness of dielectric changed from 1.4 to 2.2 mm

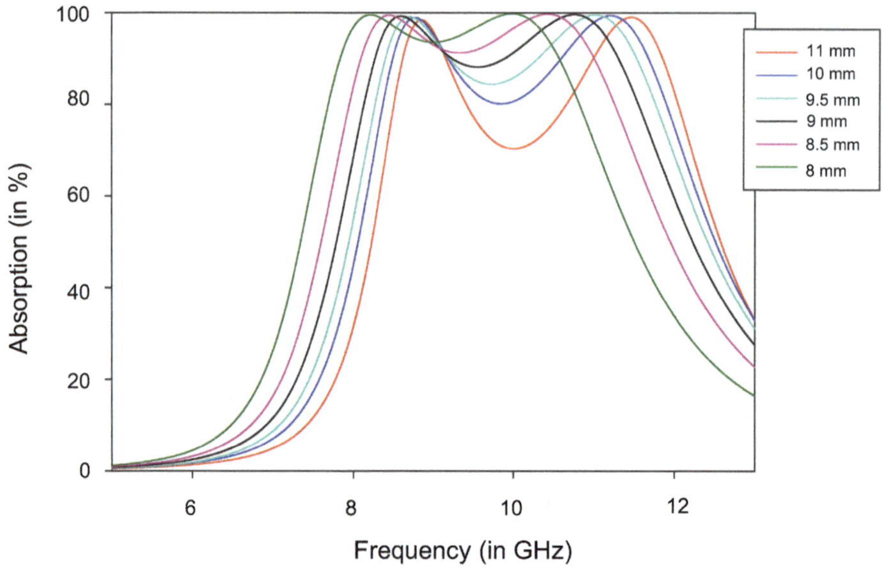

Fig. 6.17 Variation in absorption with periodicity of structure changed from 8 to 11 mm

is shifted from 9.9 to 11.46 GHz. So absorption bandwidth is enhanced at the cost of absorption efficiency.

Effect of Length of Tail Parametric study in Fig. 6.18 shows that tail length of the square pattern has an important role in determining the absorbing property of the structure and absorption bandwidth. The structure is optimized for maximum bandwidth for near-perfect absorption. When $l = 2$ mm, the real part of impedance is equal to 377 Ω for a broad frequency range of 3 GHz, and the structure acts as a good absorbing material. As tail length decreases the near-perfect absorption bandwidth is deteriorated. When l is less than 1.2 mm, the structure acts as bad absorber with reduced absorption bandwidth.

Effect of Angle of Incidence and Polarization The absorption property of the proposed structure is studied for different angle of incidence and polarization. Figure 6.23 shows the effect of different incident angles such as 0°, 20°, 40°, 60°, 80°, respectively. The absorption bandwidth remains constant for up to 50° and subsequently the bandwidth reduced. It is also to be observed that the structure loses the near-perfect absorption property when the angle of incidence is greater than 60°. The comparison regarding the effect of TM- and TE-polarized plane wave incidences is already shown in Fig. 6.19. The proposed structure shows good polarization stability for TM and TE-polarized waves for small incident angles.

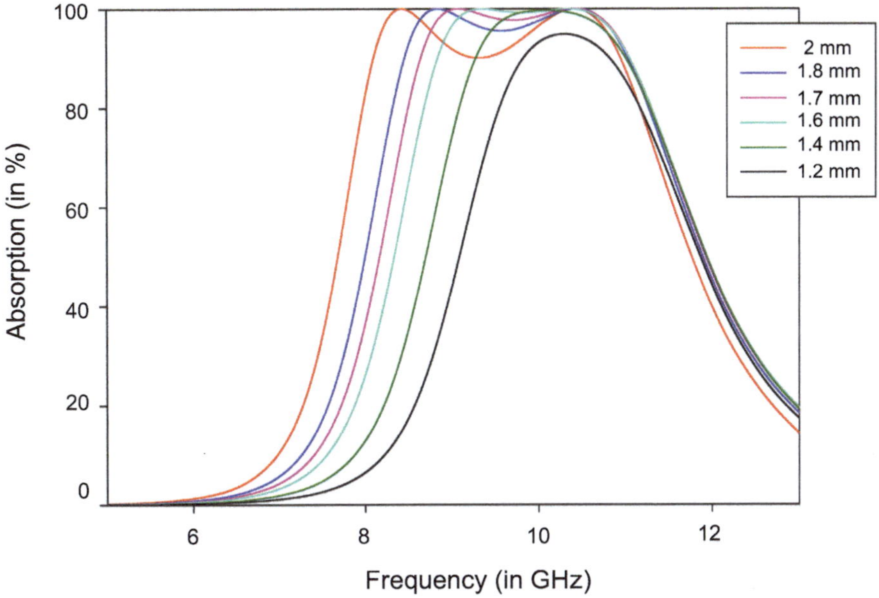

Fig. 6.18 Variation in absorption with tail length changed from 1.2 to 2 mm

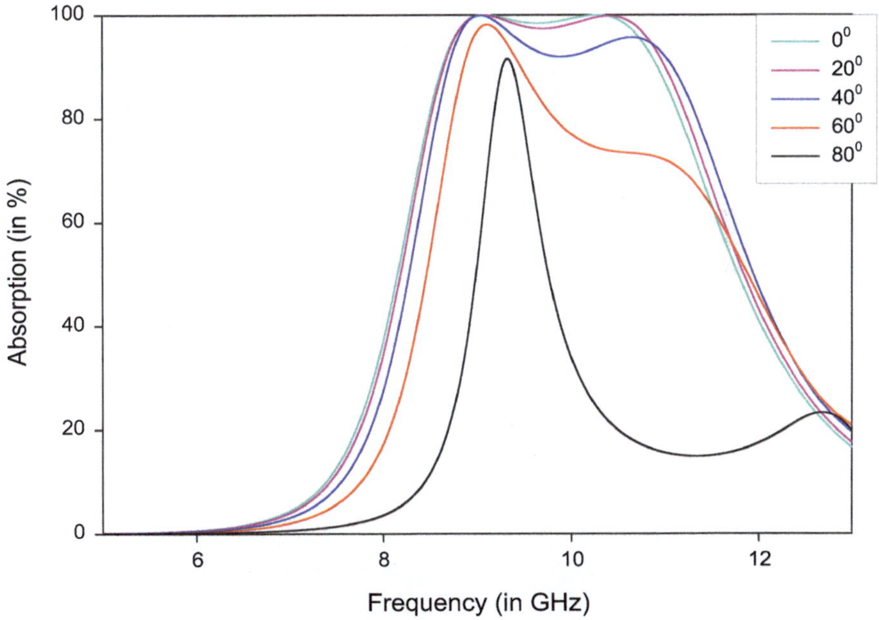

Fig. 6.19 Variation in absorption for angle of incidence changed from 0° to 80° in steps of 20°

6.6 Measurement and Performance Analysis of RAS

For experimentally investigating the absorption characteristics of proposed metamaterial absorber structures, test samples have been fabricated. The unit cell structures are fabricated with a footprint of 300 mm × 300 mm, on commercially available 1.6 mm thick FR4 substrate using print circuit board technique. Figure 6.20 depicts the photograph of experimental setup for measurement of reflection along with fabricated test sample. The experimental setup consists of two horn antennas with wideband characteristics; one for transmitting and other for receiving. The horn antennas are connected to a vector network analyzer, Rhode & Schwarz network analyzer ZVB 20 for measuring received EM wave. The transmitter antenna is placed on a wooden carriage and is positioned close to wooden arch. The fabricated test sample is mounted on a turn table and is kept at the center of the arch. In order to eliminate the near-field effects, the distance between antennas and test samples are maintained more than D^2/λ_0, where D is the largest transverse dimension of antenna's aperture. For measurement of reflected power from test sample, the receiving antenna is rotated through the arch. The measurement is taken in anechoic chamber, to suppress the effects of external fields. This arch method is used for study of bi-static RCS characteristics.

The transmitter antenna is set for normal incidence, at the bisector of arch, and receiver antenna is moved to either side of bisector. The angle is positive to the right of bisector and is negative to left of bisector. While receiving antenna is moved along arch in steps of 5° and at each position, S_{21} as a function of frequency is measured.

Figure 6.21 shows the photograph of the designed pentagon structure. Initially, we measure the reflections from a copper sheet with same dimensions as test sample, and the response obtained is taken as the standard reference for comparing the reflection characteristics of it.

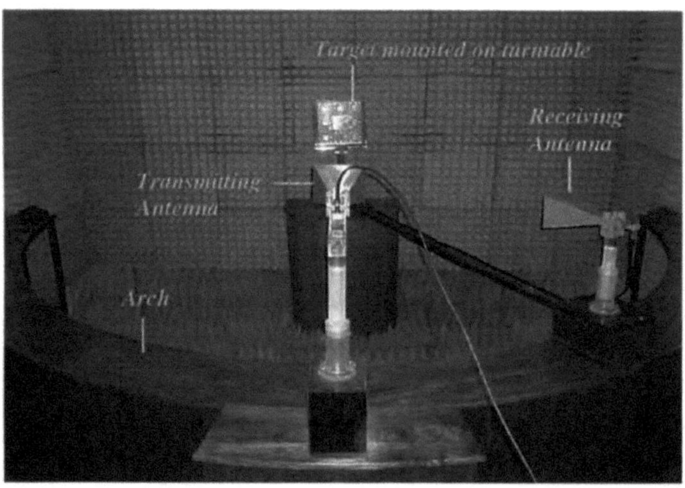

Fig. 6.20 Setup for measurement of back-scattered power

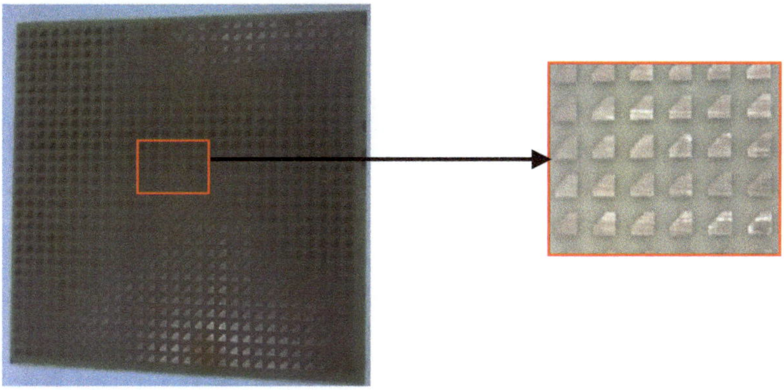

Fig. 6.21 Photograph of the designed pentagon structure test sample

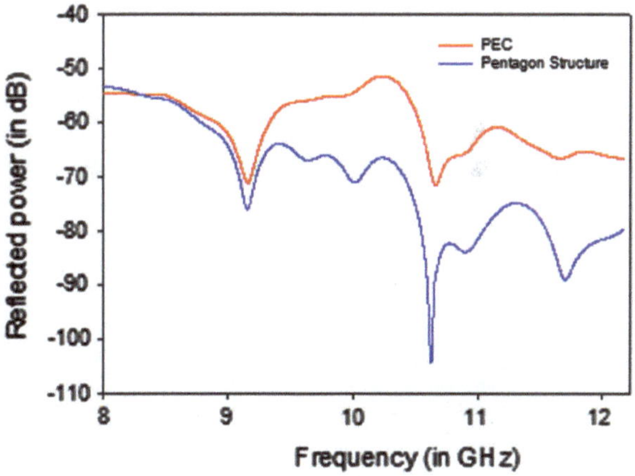

Fig. 6.22 Back-scattered signals from PEC and pentagon structure in X band

Figure 6.22 shows the scattered power from PEC sheet and pentagon structure test sample when TE-polarized EM wave incident on it normally in the entire X band. The discrepancy in reflection characteristics between copper sheet and metamaterial sheet can be considered as the absorption. The bi-static RCS in all other directions are also measured and are depicted in Fig. 6.23 along with corresponding copper sheet characteristic. It is clear from results that, scattered powers in other directions are small and it ensures the absorbing nature of the proposed structure.

The photograph of fabricated test sample of the designed novel structure is shown in Fig. 6.24. Figure 6.25 shows the scattered power from PEC sheet and Novel structure test sample when TE-polarized EM wave incident on it normally in the entire X band.

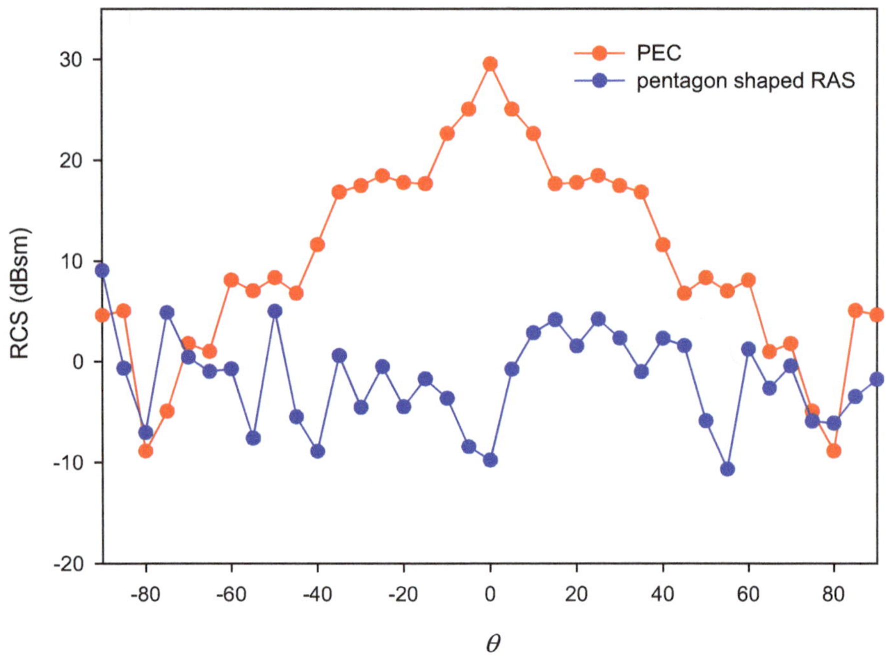

Fig. 6.23 Back-scattered signals from PEC and pentagon structure in different direction for normal incidence

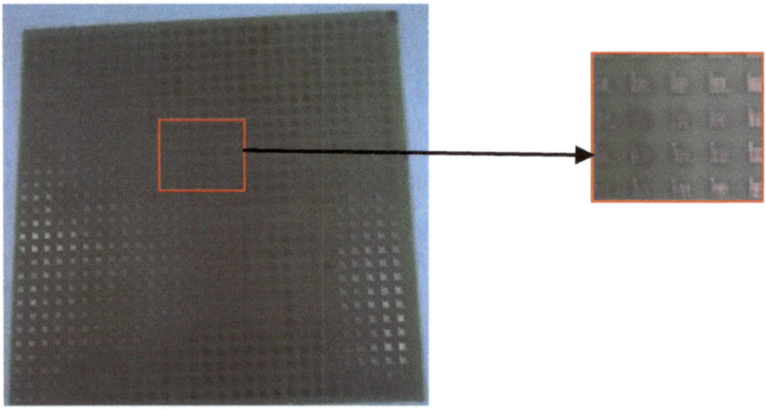

Fig. 6.24 Photograph of the designed Novel structure test sample

The discrepancy in reflection characteristics between copper sheet and metamaterial sheet can be considered as the absorption. The bi-static RCS in all other directions are also measured and are depicted in Fig. 6.26 along with corresponding copper sheet characteristic. It is clear from results that scattered powers in other directions are small, and it ensures the absorbing nature of the proposed structure.

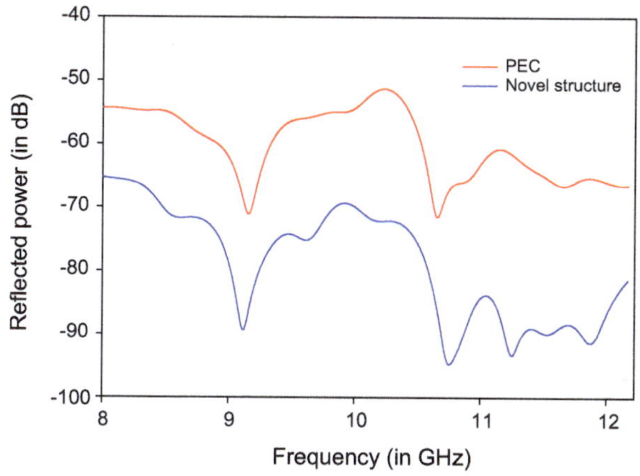

Fig. 6.25 Back-scattered signals from PEC and Novel structure in X band

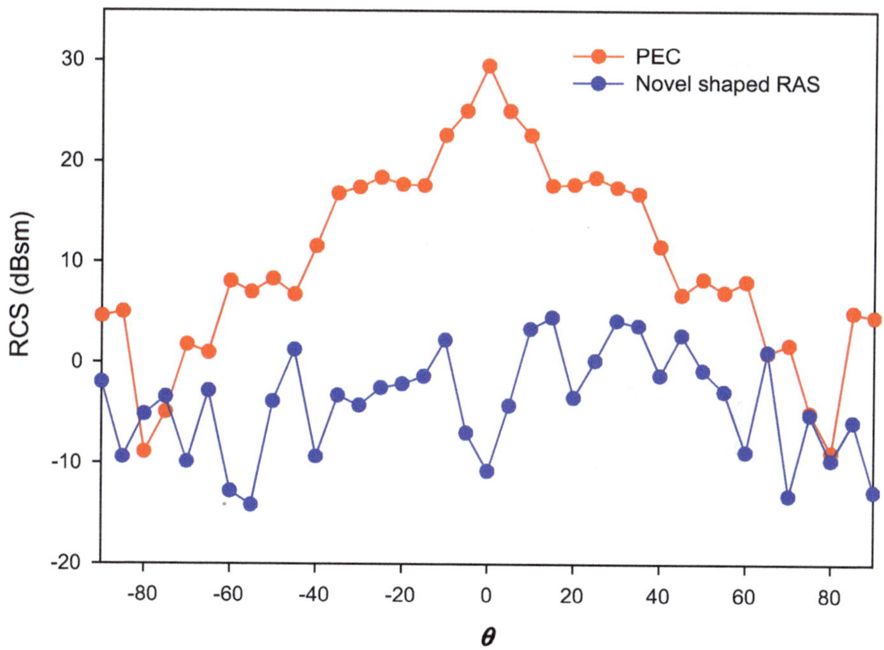

Fig. 6.26 Back-scattered signals from PEC and Novel structure in different direction for normal incidence

The measurement results show that, the electromagnetic wave in the X band, incident on the absorber structures has been absorbed and negligible power is scattered in all other directions. This reveals the absorptive nature of the proposed metamaterial based radar absorbing structure.

6.7 Conclusion

In this chapter, two broadband metamaterial structures, pentagon-shaped metamaterial structure and a novel ultra-thin metamaterial absorber for microwave frequency regime have been designed and fabricated. The broadband metamaterial absorber comprises of special resonant structures, with multiple resonances in the near frequencies. The optimum design for the structure has been achieved through soft-computing-based computational engine. The numerical results obtained from simulations shows that the pentagon structure exhibit more than 90% absorption for 2.3 GHz bandwidth. Novel structure exhibits near-perfect absorption in the frequency range from 8 to 11 GHz. Two absorption peaks are obtained at 8.43 and 10.42 GHz with absorption of 99.99%. Further the analysis of electric field distribution and surface current distribution are carried out, in order to understand the resonance mechanism of the structures. The experimental results for fabricated test sample also ensure the near-perfect absorption for a wide frequency band.

References

1. P. Saville, *Review of Radar Absorbing Materials*, pp. 19–22. Defense R and D Canada Atlantic, Dartmouth, Ottawa, Canada (2005).
2. K. J. Vinoy and R. M. Jha, *Radar Absorbing Materials from Theory to Design and Characterization*, Kluwer Academic Publishers, Boston, ISBN 0-7923-9753-3, 1996.
3. C. Caloz, *Electromagnetic Metamaterials: Transmission Line Theory and Microwave Applications*, John Wiley & Sons, Inc. Hoboken, New Jersey, ISBN-10: 0-471-669857, 350p, 2006.
4. E. F. Knott, J. F. Shaeffer and M. T. Tuley, *Radar Cross-section*, 2nd edn., pp. 64–68. SciTECH, Raleigh, NC, USA (2004).
5. Veselago, V.G., "*The electrodynamics of substances with simultaneously negative values of e and m,*" Soviet Physics Uspekhi, vol. 10, no. 4, pp. 509–514, Jan.–Feb. 1968.
6. T. J. Cui, D.R. Smith, and R. Liu, *Metamaterial Theory and Design*, Springer, NewYork, ISBN: 978-1-4419-0573-4, 367p., 2010.
7. D. R. Smith, J. B. Pendry, and M. C. K. Wiltshire, "Metamaterials and negative refractive index," *Science*, vol. 305, pp. 788–792, Aug. 2000.
8. R. A. Shelby, D. R. Smith, S. Schultz, "Experimental Verification of a Negative Index of Refraction," *Science*, vol 292, pp. 77-79, Apr. 2001.
9. R. C. Jain, "Artificial dielectrics and some of their applications," *IETE Journal of Education*, vol. 47, no. 2, pp. 67–77, Mar. 2006.
10. Y. Pang, H. Cheng, Y. Zhou, Z. Li and J. Wang, "Ultrathin and broadband high impedance surface absorbers based on metamaterial substrates," *Optics Express*, vol. 20, no. 11, pp. 12515–12520, May. 2012.
11. J. Yang and Z. Shen, "A thin and broadband absorber using double-square loops," *IEEE Antennas and Wireless Propagation Letters*, vol. 6, pp. 388–391, Dec. 2007.
12. H. Wakatsuchi, S. Greedy, C. Christopoulos and J. Paul, "Customised broadband metamaterial absorbers for arbitrary polarisation," *Optics Express*, vol. 18, no. 21, pp. 22187–22198, Aug. 2010.

13. J. Grant, Y. Ma, S. Saha, A. Khalid and D.R.S. Cumming, "Polarization insensitive, broadband terahertz metamaterial absorber," *Optics Letters*, vol. 36, no. 17, pp. 3476–3478, 2011.
14. Y. Liu, S. Gu, C. Luo and X. Zhao, "Ultra-thin broadband metamaterial absorber," *Applied Physics A*, vol. 108, no. 1, pp. 19–24, Apr. 2012.
15. W. Yuan and Y. Cheng, "Low-frequency and broadband metamaterial absorber based on lumped elements: design, characterization and experiment," *Applied Physics A*, vol. 117, no.4, pp. 1915–1921, Dec. 2014.
16. N. Jin, "Particle Swarm Optimization In Engineering Electromagnetics", UMI Dissertation Services, 2008.
17. S. Chamaani, S. A. Mirtaheri, M. Teshnehlab, and M. A. Shooredeli, "Modified multi-objective particle swarm optimization for electromagnetic absorber design," *Proceedings of Asia Pacific Conference on Applied Electromagnetics*, Dec. 2007.
18. S. Cui, and D. S. Weile, "Application of a parallel particle swarm optimization scheme to the design of electromagnetic absorbers," *IEEE Transactions on Antennas and Propagation*, vol. 53, pp. 3614–3624, Nov. 2005.
19. B. Y. Wang, S. B. Liu, B. R. Bian, Z. W. Mao, X. C. Liu, B. Ma, and L. Chen, "A novel ultrathin and broadband microwave metamaterial absorber," *Journal of Applied Physics 116*, Sep. 2014.
20. D. R. Smith, D. C. Vier, T. Koschny, and C. M. Soukoulis, "Electromagnetic parameter retrieval from inhomogeneous metamaterials," *Physical Review*, vol. 71, pp. 036617-1 - 036617-11, Mar. 2005.

FSC
www.fsc.org
MIX
Papier | Fördert
gute Waldnutzung
FSC® C083411

Kolibri 360 GmbH
Ferdinand-Jühlke-Straße 7
99095 Erfurt, Deutschland
produktsicherheit@kolibri360.de